History from the Farm

History from the Farm

edited by

W. G. HOSKINS

FABER AND FABER

London

First published in 1970
by Faber and Faber Limited
24 Russell Square London WC1
Printed in Great Britain by
Latimer Trend & Co Ltd Plymouth
All rights reserved

ISBN 0 571 09437 6

Contents

Illustrations

With the exception of the photographs of Lake Farm, Wilstrop Hall and Brick House Farm, which were produced by Mrs. Freda Wilkinson, Mrs. Trudy Blacker and Professor Michael Wise respectively, the photographs are reproduced by permission of the *Farmers Weekly*

Maps and Plans

Preface

This book had its origin in a competition for the best farm histories, organized by the *Farmers Weekly* during the year 1968. Twelve of the eighteen farms covered in the book were published as prize-winning entries. To these I have added six others in order to give a fuller picture of farms from other parts of the country. Four of these were submitted for the competition but were too long to publish in the *Farmers Weekly*, and I have edited them for publication here with the kind permission of the authors. The remaining two were a study of a Black Country Farm by Professor Michael Wise, originally published in an American journal (*Agricultural History*) in 1953; and a hitherto unpublished study of a Dartmoor farm of the long-house type by Mrs. Freda Wilkinson. I am indebted to both these authors for their ready permission to publish their essays in this book. They add greatly to its value as they are types of farms which would not otherwise have been represented.

The eighteen farms represented in this book range in location from the extreme north of Scotland (Summerbank Farm in Caithness) down to the extreme west of Cornwall (Glebe Farm in Sancreed parish). Nevertheless, it must be said at once that the coverage is uneven in relation to Britain as a whole. An overwhelming preponderance of entries for the competition came from the North and the West of Britain, practically nothing from the big arable farms of Eastern England. This distribution is not surprising, at least to me. I have always found that farmers in the West Country and the North Country are more interested in the history of their farmsteads and farms, partly because these have not changed their old identity so much as the large-scale arable farms of the eastern side of Britain. It is here that the most sweeping changes are taking place in British farming, such as the throwing-together of farms into huge food-factories, and the wholesale removal of old hedges and boundaries for the benefit of large machinery. Perhaps, too, more new men have moved in, for whom the 'back-history' of their property is not of the slightest interest. As an historian I deplore this lack of a sense of the past, but there is nothing one can do about it.

What is more surprising is that Wales produced very few entries, and those few were not characteristic, nor good enough to publish. I regret this greatly as Welsh history and Welsh farming have a unique flavour, but there it is.

These eighteen studies are not, of course, comprehensive histories of the farms concerned. This would have been impossible in the space allowed. Authors were left free to write about the historical aspects of their particular farms in any way they liked, so that what we have here is a composite picture which ranges from the prehistoric to the

11

latest improvements of the mid-twentieth century. Most types of farming seem to be reasonably well represented except the large-scale arable as said above, ranging from a smallholding of a dozen acres to mixed farms of five hundred acres or more. One hill farm in the Southern Uplands of Scotland runs to nearly twelve hundred acres.

The form of the book is as follows: after a general introduction by myself on Farms and History, the eighteen selected farms are grouped roughly by regions. The respective authors say their piece, but I have added to each my own remarks, which are designed either to bring out some points more sharply or to bring in fresh material which I thought of particular relevance. As far as possible each farm has been illustrated with at least one map or plan, and a photograph.

It must be emphasized once again that these are not intended to be complete histories, nor is the completed book intended to be a history of British farming—far from it. But it does illustrate in a detailed way many of the social and technical changes that have taken place in farming in this country, not only on the farm but in the house and its contents. There is a good deal of social history embedded in these pages.

It only remains to thank the authors of these essays for their co-operation in making what I feel is a unique book, and Miss Anne Greenwood, the Magazine Editor of the *Farmers Weekly* for her willing help on the organizing side, especially perhaps in producing the necessary maps, plans and photographs. My special thanks are due to Mrs. Jeanne-Marie Stanton for her skilled assistance over the remarkable map of Brick House Farm in the Black Country (Figure 13); and to Mrs. Susan Digby Firth who has once again typed my manuscript and with a sharp secretarial eye saved me from error and inconsistencies.

Exeter
18th July 1969

W. G. HOSKINS

Introduction

Farms and History

There are something like 400,000 farm-holdings in Great Britain, but as a farm may consist—and often does nowadays—of more than one holding, the number of 'farms' is rather less. It all depends, too, on how one defines a farm; but there are some 300,000 full-time farmers in Britain and about 55,000 part-time. So the total number of farms with a separate history is somewhere between 350,000 and 400,000. Possibly if one could count all those farms that have been absorbed into neighbouring units since medieval times we should have close on half a million histories to cope with.

Four out of five farms are still small, that is less than 100 acres, even though the process of 'engrossing', as historians call it, has been going on actively for at least 500 years. The simple explanation of this apparent contradiction is that the majority of medieval farms were only some 20 or 30 acres; a small proportion were something up to 100 acres, and a very few—generally the home-farm of the lord of the manor (the demesne farm)—ran to perhaps 300–500 acres. Yet even today the number of farms of over 300 acres is only 4 per cent of the whole.

There is in Britain a fundamental geographical and perhaps even cultural division between the North and West, and the East and South. A dividing line running roughly from the mouth of the Tees on the north-east coast of England, to the mouth of the Exe in the south-west, separates these two different Britains. On one side is the so-called Highland Zone of Britain, on the other the so-called Lowland Zone. All kinds of differences flow from this difference of topography and geology and rainfall. To the north and west we have predominantly pastoral farming, to the east and south basically arable. This deep division has survived the changes imposed by major wars, though if one looked at particular districts in detail there would be many exceptions.

Villages and isolated farms

There is one other major distinction, too, which is important for the study of individual farms. Again there are innumerable exceptions, when one looks at small districts in detail (but they can usually be explained if one goes into their history); but in the Lowland Zone the majority of farmhouses tend to be situated in *villages*, while in the Highland Zone they tend to be isolated, to stand completely alone or at most to be grouped in pairs or threes in a hamlet. Large villages are rarer on this side of Britain, whereas isolated farms are rarer in the east and south. This profound distinction is of great age historically. It may even be rooted in prehistoric times: arable farming required people to be clustered in villages in order to share their limited capital equipment (especially the eight-ox plough-teams) while pastoral farming, requiring little or no capital equipment, allowed people to settle wherever water-supply, a sheltered aspect, and suitable soils encouraged the pioneer to erect his first hovel.

But living on the village street and engaged in the co-operative farming of the open fields, or living in some lonely combe or on a hillside amid enclosed pastures, also produced two entirely different ways of life. I have quoted in another book the remark of a great French geographer: 'Between these two great facets of human settlement lie profound differences of rural civilization. It is a matter of very ancient ways of life, rooted long ago.'

Pursuing this point, perhaps I may repeat what I said in *Local History in England* a few years ago:

'Farmhouses are memorials of a peasant culture which has now disappeared. They are its monuments, just as much as those which express the culture of prehistoric times, and they are eloquent of a history which is scarcely, if at all, written in documents . . . In studying such houses . . . one should begin by noting the details of the site. The kind of soil, the shelter afforded, and the whereabouts of the water supply are all elements in the choice of site. Even where farmhouses lie in the village streets and not isolated in their own fields, considerations of soil and shelter will still operate to some extent, and the water supply will be as important as anywhere. In other words, we have to ask ourselves why a house was built just there and not somewhere else. We should also note the relationship of the dwelling-house to the rest of the farmstead . . . The essential problem for the peasant in building his farmstead was to establish a relationship (though not of course consciously) between men, beasts, and goods. The peasant does not withdraw from his cattle: he wants to keep them near, under his eye, and to keep near him under one roof all that belongs to him. This gives us what the French call the *maison bloc* or in Britain the "long-house", which groups all the essential parts of the farmstead under one roof. This is the most simple and economical of all forms of the peasant house. It is found in all regions of small-scale farming, above all perhaps in pastoral regions, but in England it is rare to find this primaeval type of long-house still surviving. It may, however, be seen in certain places around the edge of Dartmoor, though the cattle no longer inhabit the lower end of the house.'

I have included one such house—Lake Farm, on Dartmoor—in this book as a record, together with plans of the house and the lay-out of the whole farmstead. And I have heard of Dartmoor farms where the cattle are still brought into the old shippon in a bad winter, where it has been left intact, though in most farmhouses the shippon was converted into an extra parlour a century or two ago, and the cattle now have to winter in outside buildings.

Reverting to the site of the farmstead for a moment, a sheltered aspect was sometimes put first, especially in a wet and windy climate as on the western side of Britain. Water was naturally not far away, but it could be fetched (as at Carnpessack, another western farm included in this book) even from a distance: but aspect was considered to be more important. The water could come from streams (no shortage of these in the Highland Zone) or from a copious spring gushing from a hillside. The farmhouse was often built to back into the hillside, and the spring poured the living water into a trough in the back courtyard. I have stood in many an ancient farmstead and seen such never-failing springs, the fountain and origin of a house which has stood on this site since Anglo-Saxon times.

Often, too, a farmstead was situated, where possible, where two contrasting types of soil met, so that one had the fertile well-drained arable on one side (preferably to the south) and the poorer soils of the rough pastures behind. Such a farm, recorded by name in Domesday Book, was Loosedon Barton in mid-Devon, pulled down only a few years ago.

In village country it was basically the community which chose the site, usually on a light and porous soil overlying an impermeable clay or rock so that the life-giving water was trapped a few feet down and could either be tapped by shallow wells or at the springs round the perimeter of the village where the two kinds of soil—permeable and impermeable—met and produced a 'spring-line'. Aspect was probably less important in a village, which in a way produced its own shelter.

The antiquity of farms

A great number of farms in England, and possibly in Scotland too, are no more than 200 or so years old so far as their present buildings and field-arrangements go. They were created in their present form by the parliamentary enclosure movement of the late eighteenth century and early nineteenth, roughly between 1750 and 1850, or by great landlords rationalizing their estates for the purpose of better farming. Others date from earlier enclosure movements in Tudor times or in the seventeenth century, when a good deal of land was converted from open field arable to enclosed fields, often for pasture.

But in the western and northern parts of Britain, and above all in the west of England and the Welsh borderland, a high proportion of farms go back to Old English times, i.e. before the Norman Conquest. A great number also acquired their present identity in the generations that followed the Conquest, when a rising population compelled men to leave the ancestral villages and farms and make new fields and farms out of woodland, moorland, and heath.

It is probable, for example, that in a county like Devon about two-thirds of the present farms have existed since before 1066. A great number are actually named in Domesday Book and an even greater number can be identified beyond any reasonable doubt even though they are not named individually. It is generally the demesne farms (the lords' farms) that are named specifically, but the farms of their villeins can be identified also in a great number of cases. Nearly always they are to be found on the better soils of a district. Our Anglo-Saxon ancestors had a very good eye for the quality of soils and this, combined with an adequate water supply and a good aspect, led them to create their first farms. It seems certain, too, that they got this eye for attractive soils, and soils to avoid, from a traditional observation of plant-life and vegetation generally. No doubt they occasionally made mistakes, and like doctors' mistakes these were buried long ago in the earth.

Once settled, such farms continued without a break, season after season, regardless of military conquests and changes of landlord. The way we were taught history at school either said or implied that with every conquest—Roman, Anglo-Saxon, Norman—there was a clean sweep, that the past was erased and we all started again. Whatever happened to the Top People, it only requires a bit of earthy common sense to see that farming went on without interruption. Landlords may have changed, for better or for worse, but the freeman-farmer or the villein-farmer sowed, harvested, and tended his stock without ceasing. Life was chancy enough as it was. The harvest could fail in any year, perhaps for two or three years running; imported food was unknown or at the best unlikely. Even landlords had to eat, and the last thing they would do would be to wreck their own food-supplies. I think the English farmer survived all these conquests, provided he did not stand in the path of some avenging maniac like William the Conqueror in Yorkshire, who laid waste so much of that countryside in the years 1069–70. The most eminent historian of this period has said that 'from the eleventh century onwards historians have noted the sustained ferocity with which the king set his men to destroy the means of life in northern England.' Parts of northern England, extending over several counties, were still derelict seventeen years later, the object being to ensure that neither Mercia nor Northumbria should ever rise again against the Norman yoke.

Men must eat and drink before all else: and the food-grower—the farmer—is the foundation of human society. You do not destroy the farmer unless you wish to destroy the whole of society.

Other parts of Britain—not least Scotland and Wales—suffered from time to time like this, but even the most devilish of military madmen could not destroy everything even if they wished to. Even they had to eat. So a great number of ancient farms escaped these ravagings, especially those on the western side of the country. Even the Norman Conquest meant no more for most farmers, who were not what we should call owner-occupiers anyway, than swopping a Saxon for a Norman landlord. For most even this made little difference. It was rather like the bombing of the last war. If you were unlucky you were hit and lost everything; but most people escaped direct hits and life returned to normal. So for the Saxon farmer, provided he was not in the path of the mad soldiery, the weather and the harvest still mattered most of all.

Many farms, then, survived the Norman Conquest. A change of landlord, that was all. It was a massive change of landownership, more so in the top ranks than farther down the social scale. The same is true of the Great Plunder of the monasteries by Henry VIII in 1536–9, when it is estimated that between a quarter and a third of all the land in England and Wales changed hands. How much this affected the great mass of tenant farmers, always the majority, is still an open question. Most of the expropriated land went to the old families in each district, who were not slow to profit by the dissolution of the local monastery; and some went to the New Rich, especially the high Tudor officials, who carried through the administrative side of the transfer and feathered their own nests richly. Again, it was a matter of chance whether the tenant farmer got a better landlord than his former monastic landlord, or a worse one. There is plenty of evidence that some of the New Rich regarded their acquisitions purely as investments—rather like a take-over bid of an under-developed company today—and screwed the utmost out of their new tenants. But it is equally likely, in my opinion, that they behaved correctly and tried to avoid a reputation for rapacity. After all, they were newcomers to an old landed society and, except for a minority, would be more likely to feel their way carefully, following the traditional code of the countryside in the hope of being accepted socially.

Continuity

If so many farms, in southern England at least, survived the Norman Conquest without any break in their life, the question then becomes—and it is much more difficult to answer—what happened at earlier conquests, the Old English Conquest, and even the Roman Conquest? Is it possible to think of a minority of farms at least going on without a break in these most distant periods of time?

There are no documents to help us in this search for continuity, except an occasional reference in an Anglo-Saxon land charter. Nor is there any direct archaeological evidence as yet, except that of air-photographs which show up ancient field-systems in various parts of the English countryside where farming is still being carried on. But we have not begun to date these field systems at all accurately. Archaeologists often referred to them as 'Celtic fields' but even this is now being cautiously dropped in favour of a non-committal 'ancient'. As for farmhouses and buildings of the pre-Saxon period, they may well lie underneath the buildings in use today. Until an existing farmstead site is vacated and then completely excavated by experts, level below level, we shall never know how old such a site is. When, for example, Loosedon Barton in mid-Devon, to which reference has already been made, was demolished a few years ago, it was a house of about 1600. But we know it was mentioned by name in Domesday Book, so where was the farmhouse of the eleventh century? Was it underneath the house of 1600, and underneath other rebuildings, or was it elsewhere in the farmyard? Perhaps only a few yards away. If only one could have got that site excavated at the time!

There is one other suggestive kind of archaeological evidence, however, and that is where a modern farmstead stands inside or adjacent to an early earthwork of a non-military (i.e. civilian) type. Such a farmstead is Bury Barton, also in mid-Devon, where the existing buildings—many of them dating from late medieval times—are still grouped in the corner of a well-marked rectangular earthwork which was probably a

19

Romano-British farmstead with modest defences. Bury, too, is mentioned in Domesday Book, and though all this is a long way from proving unbroken continuity from the Iron Age to the present day, it is highly suggestive if one accepts the argument that in many parts of England which escaped the first impact of military conquest there was no reason ever to abandon a good site.

In the following book I have ventured on very slender evidence to suggest that Bartindale Farm on the east Yorkshire wolds may have had its origin in a large Iron Age farm; and down in Dorset I have a shrewd suspicion that Liberty Farm in Halstock parish may have been farmed as part of a Roman villa estate before the Saxons took it over.

Air-photographs are increasingly revealing widespread evidence of farming along the gravel terraces which flank many rivers in the Midlands and elsewhere. This is very evident in the valleys of the middle and upper Thames, the Warwickshire Avon, the Welland and its sister rivers in the east Midlands, and in the west Midlands around Pershore and Evesham.

Although at the moment British historians are very reluctant to think in terms of such a long continuity of farming, I myself think that in twenty years' time it will be a proven fact in some favoured parts of the country. By then it may well become a commonplace of agrarian history as it has long been in France. I think we shall have proof of the continuity of the Romano-British villa estate into later history and down to the present day in a superficially changed form. Indeed, in one fertile piece of the middle Exe Valley just north of Exeter, on the river gravel terrace, air-photographs have revealed extensive traces of 'ancient' field-systems, perhaps one overlying another of a different period; while only two or three miles away is a group of some twenty Bronze Age burial mounds, the burial places of what we might roughly call the local squires of the time. The mere fact of so many burial mounds raised the question at once: Where did these people farm? And I think the answer is given by the ancient field patterns on the gravel terrace by the river. If these are Middle Bronze Age in date, as some of the excavated barrows would suggest, then farming may well have been carried on continuously on these rich soils for something approaching 4,000 years. I see no reason why, in the far south-west, this farming should ever have been interrupted.

Later farms and boundaries

So far I have discussed mostly the oldest kinds of farms, and some of the problems connected with their history. But thousands of other farms came into existence after the Norman Conquest, in the generations between about 1100 and 1350 when population was rising steadily and compelling younger sons and daughters to clear new land from the surrounding woods, the marshes, and the moors. Clearance of woodland was often a family enterprise, and so too was the making of a farm out of moorland and heath. The drainage of potentially rich marshland, because of the scale and nature of the work, was generally a great co-operative effort shared in by whole villages and often directed by some monastery which owned the property. Most of the Somerset levels were drained by the abbeys of Glastonbury, Muchelney, and Athelney between the tenth century and the fourteenth.

In 'village country' this expansion into the woods and heaths and moors did not usually produce new farms, but extended the boundaries of the existing open arable fields. But elsewhere, chiefly in the west and north of England, thousands of new farms appeared in the landscape in this period of expansion—until the Black Death put an end to the necessity for seeking new land. It was after 1349 that most of the desertion of villages took place. There are some 2,000 known sites of 'lost villages'. In this book we have at least two—Bartindale and Wilstrop, and perhaps Hunshelf, all in Yorkshire—but this is accidental. These deserted villages are in fact most numerous in the Midlands and eastern England.

Outside the 'village country' we also see a new grouping of farmsteads appearing in the landscape—two, three, or four farmsteads all loosely grouped to form a hamlet, with their lands spread out in a sort of fan around the settlement. In point of fact we do not really know the origin of the hamlet. Was it an aboriginal form of settlement, that is a group of farmsteads from the beginning? Or did it arise by the division of an original farm between two or three heirs, each with his own farmstead and his lands intermingled with those of his kinsmen? Where we get this intermingling of fields in a hamlet, it seems likely that it arose from the lands of a single ancient farm being divided between heirs. But in other cases the hamlet still requires a bit of explaining.

Whatever the antiquity of a farm, it had to have boundaries from the beginning.* As far as possible natural features were chosen—a stream, a well-defined line of hills, the edge of an ancient wood, or perhaps an already existing path. But where no clear natural boundary existed, one had to be made, and here we find long hedgebanks. These were originally excavated out of a ditch and the earth thrown up, and often faced with stone for greater strength. The hedge comes out of the ditch as a physical fact, so that in law to this day the ditch—if there is some trouble over blocking for example—always belongs to the owner of the hedge adjoining. In addition to perimeter hedges, or walls, forming the boundaries of one farm against another, there would also be internal hedges and walls on a farm dividing one field from another. These internal hedges and walls might be of varying dates, depending on the particular history of the farm itself.

Recently a theory has been put forward by Dr. Max Hooper (and discussed in my book *Fieldwork in Local History*) that we can date hedges by taking sample counts of the various species of trees and shrubs growing in them. We count both sides of an un-managed hedge so as to allow for difference of aspect, and we can then estimate the age

---

* Perhaps not from the very beginning. The earliest farms in a thinly-populated countryside may have been very vaguely defined for a time; but as the countryside filled up with settlers well-defined boundaries would soon have become necessary to avoid disputes.

of the hedge by allowing one species for every hundred years of the hedge's existence. There are qualifications to this simple formula, but on the whole it seems to hold true wherever it has been tested.

There have, of course, been many changes of boundary between farms, especially where one has had single large landlords who at some point in time, generally in the eighteenth century or the nineteenth (but not only then) sorted out historic anomalies between boundaries and created a more workable pattern of farms. All the same, it is surprising how often one finds farms, certainly in the West of England, which have not changed their historic boundaries, or have made such minor changes that they can easily be detected. This is another large subject which calls for much further exploration.

The most revolutionary changes of boundaries were made in the open-field parts of Britain. At varying dates, but again mostly between about 1700 and 1850, the open-field strips were enclosed, formed into blocks to produce a 'modern' field pattern, and the lands belonging to an old farmstead thereby completely reshuffled. The farmhouse and buildings may well remain in the village street, but the identity of the farm-lands only comes into existence at the time of the enclosure. Often, too, the old farmstead was abandoned (usually turned into cottages) and a new farmhouse and buildings erected in the middle of the new fields. This is why so many British farmhouses and buildings date from the late eighteenth century or the nineteenth. In most such cases, they replace a much older site on the village street.

Recent changes in British farming, chiefly as a result of increasing mechanization, have led to greater changes in boundaries than at any previous time. Farms have been thrown together to form huge arable units, and old hedges swept away to make much bigger fields for economical working. It is thought that we are losing something like 10,000 miles of hedges a year by these changes. This is not the place to start an argument whether these changes are all to the good. I myself do not think they are.

Farmstead types and plans

The position of the farmstead in relation to its fields, provided the ancient boundaries have not changed much, also tells us something about its historic development as a farm. Is it central to the fields, or does it lie well away to one side? If the latter, it suggests to me that the farmstead had originally once been more or less central, but that later expansion of the farm by the clearance of woodland, or taming of moor and heath, over many generations had left it to one side as it were.

The farmstead itself has its own special interest. By this I mean not only its position in relation to the farm as a whole, but the relationship of the house and farm-buildings to each other as a group. Nothing is accidental or haphazard: there must have been a good reason for whatever arrangement we find today, even if it is not now apparent. There are in Britain some clearly defined types of farmstead, depending basically on the kind of farming that was traditionally carried on, and of course on the scale of

22

farming. Most farmsteads form a complex of buildings; and it is possible that the use of some buildings has changed over the years, just as in an old farmhouse some rooms have changed their function with the growth of the house or an increased demand for modern comforts and conveniences. The most dramatic change in one way, or perhaps I should say the one most easily detected, is the conversion of the shippon or mistal of a 'long-house' into a parlour for family use. The animals were turned out of the house after centuries of sharing with the family—an arrangement which had benefits for both sides—and put into separate buildings in the yard. I have seen evidence of this change of use—a change in cultural ideas almost—in many a Dartmoor farmhouse.

In these changes in the interior plan of old farmhouses, women have played a major part. One sees this certainly right from the early sixteenth century when probate inventories allow one to reconstruct house-plans with a fair degree of accuracy. We can trace the growth in the number of rooms in the farmhouse, and an increasing specialization of use on both floors. Two books by Maurice Barley help to show us a good deal of these changes—*The English Farmhouse and Cottage* (1961) and *The House and Home* (1963). But a vast amount of work still remains to be done on the study of farmstead types in this country, and even more on the study of farmhouse plans through the centuries. The documentary evidence is abundant, though it is more complete in some parts of the country than others, and the inventories also vary in the amount of detail they give. A good example of the use of probate inventories for the second half of the sixteenth century—the time when so many farmhouses and cottages were rebuilt—is to be found in Michael Havinden's *Household and Farm Inventories in Oxfordshire* (1965). This includes transcripts of 259 inventories for the forty years between 1550 and 1590. Other inventories have been printed for parts of Essex, Bedfordshire (for a very limited period), Devon, Nottinghamshire, and a few from north-east England. But the number that have been put into print is only a tiny fraction of those that exist in record offices, awaiting inspection for the first time since they were written centuries ago.

I said a moment ago that women had played a major part in bringing about changes in the complexity and comfort of farmhouses. These changes have always required money, it is true, but in periods of inflation (as in the second half of the sixteenth century for example) farming profits were high, and similarly during and after most great wars. Left to themselves, men would probably effect few changes in their domestic surroundings. It was their wives, from the sixteenth century onwards (and earlier for all we know) who saw improvements in other people's houses and agitated at home for something better. The money was there, but it was they who made sure where some of it was spent. I first stumbled on this simple idea when going round Devon farmhouses immediately after the Second World War, and saw how many ancient farmhouses were being altered in order to get modern plumbing, a proper water-supply, electricity, and more air and light. It was clear from conversations on the spot that it was the wives who had agitated for these improvements while the cash was there. Once that idea was born, I could see in Elizabethan and later records how farm-wives must have been responsible for the major domestic changes in past centuries—the development of the kitchen as a separate room, freeing the hall from cooking and much messy activity, and the development of upstairs bedrooms with bigger windows and even fireplaces. Even back in early medieval times I feel sure that the development of the simple one-roomed hall-house into a two-roomed house, the second room being called the bower in most parts of Britain, was due to the need of women for more privacy. If it were not for women one sometimes feels that a great number of men would still be content to live in a cave.

The contents of the farmhouse

There are various kinds of records for tracing the development of the farmhouse plan but this is not the place to discuss them. The most valuable are the probate inventories, and at their best these also list the contents of the house room by room down to the merest trifles. The inventories run from the time of Henry VIII to George II, and tend to die out about 1730. A small number survive from after that date. In some parts of England the surviving inventories do not begin until about 1600 or even later: the earlier ones have been lost or destroyed. And they vary in the amount of detail they go into.

The most detailed inventories list all the farm stock, gear, and implements, all the crops growing or in the barns, and everything else connected with the farm down to the poultry and the dung-heap in some cases. Indoors, they list every stick of furniture, every dish, the fire-irons, domestic furnishings if any (such as carpets and curtains) and the books if any, the musical instruments, and what one might call the comforts of life. Thus I copied before the war the complete inventories of an ancestor of mine, a big yeoman farmer who died in December 1625, and that of his widow who died in 1631. Between them these two documents gave me a complete picture of their large farmhouse in east Devon and everything they possessed. From the first record I discovered that George Hoskins read the Bible 'and other little books', played the 'treble viall', wore a silver ring, and drank gratifying quantities of wine—the inventory mentions 'five hogs-heads wyne barrells with all manner of wooden vessels'. It is a splendid picture of an Elizabethan yeoman, even if sketchy.

Though these inventories generally peter out in the second quarter of the eighteenth century, one can sometimes find detailed valuations made for the purposes of a farm-sale at a later period. As an example of one of these I have selected here a long valuation made for a sale at Hunt Court Farm in Gloucestershire in March 1831. This is just as valuable as an Elizabethan probate inventory as a picture of the contents of a big farm. All such auctioneers' records should be carefully preserved, and indeed are being collected now into county record offices. Inventories made for insurance purposes today will be of equal interest in 50 or 100 years' time. One feels sometimes that no scrap of paper should ever be destroyed.

FARMS AND HISTORY

I have touched on only a few aspects of Farms and History in this brief introduction. The following eighteen farm histories fill in the details from John O'Groats in the north of Scotland down to Land's End in faraway south-western England '*with all the numberless goings-on of life.*'

Four West Country Farms

An Ancient Cornish Farm

There is so much I could say about this fascinating little farm of Carnpessack in the far west of Cornwall. Though it is recorded by name as long ago as 1230 it is almost certainly much older than this as a site. The spring of never-failing 'sweet water' possibly supplied an Iron Age farmstead. All the Iron Age and Romano-British pottery was found on the steep slope 'bottom' immediately above the spring. The Bronze Age sword was found a few feet from the spring and is no evidence of settlement—possibly a wounded soldier of 3,000 years ago went to drink at the spring, and died there. But the pottery may well indicate what the archaeologists call a habitation-site, which lasted for some centuries. At a later date—but when it is impossible to say—the farmstead was built on a more sheltered site, but the 'sweet water' still had to be got from the ancient spring.

At some unknown date, too, one of the many Cornish saints took up his abode by the spring and a small chapel was built. Not a stone now remains of this chapel, nor do we know the name of the saint, who was really a missionary in the earliest days of Cornish Christianity. Cornwall, like Wales (from where many of these early missionaries came) is full of 'holy wells' and ancient chapel-sites. There was another such site at Sancreed also (see p. 34).

Notice, too, how tiny the fields of Carnpessack are—only 56 acres, divided into 17 enclosures, an average of about 3 acres each. Several of them are under 2 acres. These small fields with their solid stone 'hedges' gave vital shelter to both stock and crops.

*

CARNPESSACK
by Rowena Bush

Carnpessack is a very humble little dairy farm of 56 acres, carved many centuries ago out of Goonhilly Downs. The two great attractions are the lovely sea views (Kibben Cove) and its long history. It is first mentioned in 1230 as Karpesek, 1300 Carpesek, 1346 Carnpesek. The name means *car* or *ker*, 'a fort', and *pesek* 'decayed or broken down', so it was called after some ruined or decayed earthwork of which there is now no trace. It was the home of one Carpyssack, a fisherman of Coverack who, during the Reformation, defied the orders of Henry VIII and kept the forbidden feast days (these were the days on which the bulk of the fish was eaten and his protests may have been more a matter of economics than religious beliefs). Lord Godolphin sent soldiers against Carpyssack and he was taken in chains to Helston, hanged, drawn and quartered in 1537. The full story is told in *Tudor Cornwall* by A. L. Rowse.

I purchased the property twelve years ago from the late Mr. William Carne Pezzack of Penzance, a descendant of the bold fisherman. Mr. Carne Pezzack was a keen historian and presented the Bronze Age sword (3,000 years old) which was found at the spring in 1926 to Truro Museum. In 1963 an electric pump was installed and during the excavations thick sherds of cordoned pottery of the Iron Age were found, also thin sherds of Romano-British pottery, together with a handle of a jug dated as sixteenth–seventeenth century. These finds show that water had been taken from this spring since Iron Age days, over 2,000 years ago.

The spring is some distance from the farmhouse, but only a short distance from the house is the quarry where the stone was dug for the original buildings. It is shown on the sketch-map at the north end of the yard. There is always some water here (rain water) and one side is sloped so that cows and pigs could be driven down to it to drink in ancient times. There is also a grassy dell by the side of the farm-road which fills up from November to April, and I understand that people washed their clothes and persons there. But all 'sweet water' for drinking, cooking etc., and of course the horses, as far as I know has always been fetched from the spring.

Mr. Carne Pessack installed a $1\frac{1}{2}$ h.p. Lister engine at the spring in 1950 and built the pump-house. Previously water was brought up in old churns by pony and float. Womenfolk I'm afraid often had to walk to the spring to fill a kettle, and one day in February 1963 an icy east wind froze all the pipes and I had to do the same.

There is nothing unusual about the spring: the water runs out through fissures in the gabbro rock into a collecting chamber. It is beautifully clear water and even in that terribly dry summer of 1959 there was ample. This spring is still the life of the farm, and although mains water has now reached the neighbourhood, my spring water costs only a fraction of what that costs.

The small cottage-type farmhouse is built of cob and stone with 3-ft. thick walls. L-shaped, it is joined to the barn, and has altered little since Carpyssack's days. A pigeon house still adjoins and in the big kitchen there is an open fireplace 7 ft. × 4 ft. with bread ovens each side. These were heated by gorse faggots. The thatch roof is replaced with slate. The stone floors have pieces of rock (gabbro) incorporated.

The land lies in 17 small fields, numerous hedges breaking the wind and sheltering the stock. I am indebted to my friend Mrs. Edith Dowson of the Cornwall Archaeological Society for her research on the field names. Lower Park Chapel and Higher Park Chapel take their names from a medieval chapel which was near by. Lower Praze means lower green (an early field). Little Park Noweth is little new field, Park Noweth new field, Park Skebow a sheltered cow field. Park in south-west England is a common word, meaning an enclosed field. *Ske* or *skew* means 'shelter' and *bow* means 'cow', as in *bowjy* meaning 'cow-house'. These are all old Cornish words.

Little and Great Park Bean are uncertain in meaning. Barrow Croft contains a fine example of Bronze Age tumulus (barrow) about 85 ft. across. Higher and Lower East Fields explain themselves. Higher, Great and Little Moor were apparently croft at one time, also Croft, the Croft and Little Croft.

The soil varies from light fertile sandy to heavy blue clay. In the yard are some fine examples of gabbro rock and the spinney has a wonderful gabbro rock formation.

There are some old deeds of 1836 in copperplate with leaden seals, showing the farm changed hands that year for £900 and was let the following year to Henry Richards, yeoman, for the sum of five shillings in coin of the realm.

FIG. 1. Carnpessack Farm, near Coverack (Cornwall)

A Farm near Land's End

The origin of this interesting farm was concealed for a long time by a rather meaningless change of name when it became an inn. It was always known as Glebe Farm before that, and has now reverted to its old name. But even if we did not know its true name, one would have suspected its origin by the way in which its fields completely surround the ancient church and settlement of Sancreed. Knowing that it was the glebe land of the vicarage then leads one to examine the documents known as glebe terriers. These were more or less detailed schedules of the house, buildings, and lands belonging to the vicar for the time being.

It seems clear from the fact that the farmhouse was not built until about 1840, and from the description of the vicarage in the seventeenth century, that until the nineteenth century the vicar farmed his own glebe lands, as was often the practice until recent times. Thus the terrier or schedule of 1672, after describing the vicarage house in some detail, goes on to speak of 'one fold for sheepe, one Mowhay for Corne, and one backe yard for fewell and other necessaryes'.

The greater part of the glebe consisted of common or rough grazing, so necessary in a pastoral economy like that of west Cornwall. There were also 25 to 30 acres of arable and pasture and meadow; the terriers vary slightly in their estimates. And this small acreage was divided into no fewer than twelve closes, averaging between 2 and 3 acres each. Such tiny fields are typical of the south-west of England, especially as one goes further west. Wind is the greatest curse of the Cornish climate except in the sheltered valleys, and small fields surrounded by massive hedgebanks of earth and granite were necessary for the protection of crops and stock. In recent years farmers have bulldozed some of these ancient hedgebanks in order to make larger fields, but there are limits to this process. One is where to put the quantities of stone so demolished, and the other is that the wind still blows as strongly as ever and frequent shelter is still important.

*

GLEBE FARM, SANCREED (CORNWALL)
by Kathleen Hawke

In the remote and tiny village of Sancreed, 6 miles from Land's End, a noble-looking farmhouse nestles under the brow of a hill. Together with 56 acres of farmland and

1. Carnpessack Farm, near Coverack

2. GLEBE FARM, SANCREED

some 90 acres of moorland it was the home of my youth, and although it is now scheduled as an historical monument both the house and the farm had previously undergone various changes.

Probably only some of the local folk are aware that this house, substantially built of finely cut granite, was the local inn until 1890 and bore the name 'Bird-in-Hand', the sign depicting a hand holding a pheasant. With its wide open frontage it stands opposite the church. It was described as 'newly built' in 1841, replacing an old thatched inn which was a long low one-storey building without a sign.

In 1897 my grandfather became the tenant of this farm, then known as Glebe Farm, and the property of the Ecclesiastical Commissioners. My father became the tenant in 1914, and seven years later the owner, with the exception of 11 acres retained by the Church Authorities. The name was then changed to 'Churchtown Farm'. The interior of the house still resembled a pub. The wide passage inside the front door was boarded to a height of 5 ft. and the floor was of large granite paving stones. The kitchen, on the right-hand side, also had a granite floor and the farther wall was occupied by a large cooking range and open chimney. In addition to a built-in coal-house there was a built-in settle. A door led into a lobby which had two more doors, one leading out to a cobbled court and the other giving access to the former wine cellar with a cobbled floor and two windows protected by iron bars. Grandmother used this room as a dairy.

Returning to the front passage, and continuing right, one came to a little room which might now be called the 'snug'. It was boarded to the ceiling and had benches all round and a granite floor. My parents had the wall adjoining the lobby removed and an oil cooker installed. The kitchen then became the dining-room, a tiled grate replacing the cooking range and open chimney. On the left-hand side of the passage, before one reached the main bar, there was a serving hatch which had been papered over. The main bar had a grate at the far end and a board ran all round the room about 4 ft. off the floor. The benches had been removed. There was a similar room at the back of this bar.

The wide staircase gave access to a long landing with five bedrooms leading off from it. Between the two front bedrooms there was a screen which was folded back when banquets were held there in the prosperous days of tin mining in the parish. My parents replaced this screen with a wall. On the second floor there was a large attic or garret as we used to call it. The doors in the house were beautifully panelled and most of the rooms had heavy shutters. Water was pumped from a well a few yards away from the front of the house until pipes were laid for a supply downstairs and upstairs round about 1922.

When my father sold the farm in 1944 the name was changed to 'Bird-in-Hand Farm' and more structural changes were made to the house. The wall dividing the two downstair rooms on the left was removed and a large window added at the back, the existing back-room window being replaced by a long narrow one. Our kitchen became the office and what was then our dining-room once more became the kitchen with a Rayburn and a horizontal window replacing the vertical one. The outside galvanized 'privy' dis-

C 33

appeared and the new one was installed in part of the old dairy. A bath with hot water took the place of our galvanized bath upstairs. Electricity was installed throughout.

A building attached at an angle on the right-hand side of the house was known to us as the trap-house and another built on to the left housed the pig foods and an open chimney where the water was boiled on pig-killing days. Being an inland parish Sancreed pigs were much sought after at the local market at Penzance, 4 miles distant. Pigs from the surrounding coastal parishes were often fed on fish offal. The inhabitants of Sancreed used to be referred to as 'Sancreed Pigs' which really was not so disparaging as it sounds. The pigs' houses were called *crows* by us, the word *crow* being the old Cornish for a shed or sty. Cornish ceased to be spoken some 200 years ago, but many words survive—sometimes in a corrupted form—in buildings, field-names, and so on. The pigs' houses were too near the house and were demolished by the new owner. A small enclosure, also near the house, was always referred to as Keel Alley, a corruption of Skittle Alley, and one of the fields was called Bowling Green, reminders of an almost forgotten era.

The farm buildings, at the back and right-hand side of the house, were somewhat scattered, with a footpath running through the spacious farmyard or 'townplace' as we called it. The farmstead was situated in the centre of the farm, and quite a number of fields had footpaths leading to the village. Trees, and the 90 acres of moorland above, provided shelter. The top of the moor was known as The Beacon, and was 564 ft. above sea-level, commanding excellent view of St. Michael's Mount and the coastline. Hence its use as a beacon in former days.

The low-lying fields were more fertile than those bordering the moorland, but the latter was essential as it enabled farmers to carry far more stock than they could otherwise have done. On the rising ground, now called Chapel Downs, the remains of an ancient chapel can be seen, with a Holy Well beside it. Such holy wells are common in Cornwall, as they are in Wales, and this one no doubt nourished the Celtic saint who dwelt there and brought the Christian faith to the natives of the district long before the parish church was built. This saint was probably called Sancredus, but nothing is known about him, unlike many of the Cornish saints whose 'Lives' were recorded many centuries ago. The well was dedicated to St. Sancredus, and its unfailing flow of fresh water was used by my grandfather to pipe a supply to his fields. It is nice to think of the saint's spring being put to use within living memory.

There are some curious field-names on the farm, such as Higher and Lower Conuffin, and the three Ouse Ouths. The old glebe terriers throw some light on these, as they give us the names of the fields some 300 years ago.

The fact that the farm was called Glebe Farm, and was once owned by the Ecclesiastical Commissioners, gave me a clue as to its origin. It had evidently once been the glebe farm of Sancreed vicarage, and an enquiry at the Cornwall Record Office in Truro produced five documents—known as 'terriers' or schedules of lands belonging to the parson—describing the glebe. The earliest was undated but went back to the first years of the seventeenth century, the next was dated 1613, and the others were 1672, 1727, and 1746. The terrier for 1672 was very full, and gave a detailed description of the vicarage house, and of all the fields belonging to the glebe. What we now call Higher and Lower Conuffin were then called Great Gonoughan and Little Gonoughan, but what this name means I still do not know. It is evidently Old Cornish. The three Ouse Ouths appear by the Cornish word of *hewas* in 1672, which means a 'summer field'. This probably means that they were shut up for summer hay while the cattle grazed on the open moorland above.

In 1958 Bird-in-Hand Farm changed hands again. Three fields had already been sold for development, a neighbouring farmer bought three more, and another acquired four fields. And the farm, having undergone many changes, has now reverted to its ancient name of Glebe Farm. These fields have been farmed from time immemorial.

A West Devon Farm

Bratton Clovelly is a large parish in the wilds of West Devon. Most of it stands high, 500 to 800 ft. above the sea, on poorish soils. Chimsworthy is an isolated farm, typical of the majority in Devon, far out in the parish. Though it is not recorded by name until 1286 it must have been one of the fifteen farms recorded under the manor of Bratton Clovelly in Domesday Book (1086).

It is possible that farming has gone on here much longer than this as the parish of Bratton Clovelly has at least three farms whose names go back to Celtic times—Breazle, Maindea, and Wrixhill. Breazle was indeed the home of a chieftain, as the name includes the Celtic word *lis*, meaning 'a court, a palace'—more probably a sort of local squire in Celtic times. In modern language it was the squire's or chieftain's dwelling.

Breazle is only a mile and a half as the crow flies from Chimsworthy, so it is possible that the latter was part of the British chieftain's estate. At a later period it came into the hands of a Saxon owner (Cemmi) who gave his name to the farm, which has survived to this day.

It is likely that the incoming Saxons in the seventh and eighth centuries drove the Celtic natives into the poorer land, the yellow-clay uplands: hence the survival of this group of pre-English names in this part of Devon.

The next parish to Chimsworthy, again only a mile and a half away, is called Germansweek, taking its name from St. Germanus of Auxerre, to whom the parish church is dedicated. He was a missionary-saint who visited Britain in 429 and again in 447. This early dedication at Germansweek may well indicate the foundation of a church here as early as the middle of the fifth century, and that in turn implies the survival of a sufficient rural population to justify the setting-up of a Christian church in this remote part of Devon.

It is also very significant that the inhabitants of Germansweek are of a distinctly 'Irish type', dark-haired and high cheek-boned, unlike those of the 'Saxon' villages near by. This ancient Celtic type still goes on and on, so great is the power of heredity—vastly underrated in these days when we think that environment is everything, and that if we change the environment we change everything. Still, every farmer knows the profound importance of heredity and the persistence of old strains in animals, and so it is with men and women.

I do not know what the physical characteristics are of the Bratton Clovelly people, whether they resemble those of Germansweek on their western border, but I have no doubt that in some parts of the parish farming has gone on since Celtic times; and I

36

would expect to find a considerable element of a dark Celtic type among the older inhabitants, those families who have lived for a long time in the parish.

*

CHIMSWORTHY, BRATTON CLOVELLY
by Sybil Tope

Chimsworthy is a west Devon farm of nearly 200 acres. It is north of the A 30 and south of the B 3218 roads, 9 miles from Okehampton and about 14 each from Holsworthy, Launceston and Tavistock. Bratton Clovelly, the nearest village, is just over 1 mile away.

The original water supply was from an open well set in the shillit at the top of the yard. It is at the foot of a high bank on top of which is the kitchen garden. Water drips in from above and at the sides from a spring, and the well never runs dry. Even in the coldest weather there is never more than a very thin layer of ice on it.

Outside the kitchen window at the east side of the house is a well from which water used to be hand-pumped into a sink in the corner of the kitchen. This mode of getting water into the house succeeded that of carrying it in buckets from the open well until 1960 when water was piped throughout the house and buildings from a borehole sunk in Great Meadow.

Chimsworthy is south-facing and sheltered to the north and east by Broadbury, the highest ridge between Dartmoor and the Atlantic. With three outside doors opening directly into the rooms, the house is inclined to be draughty, particularly when the wind is in the east, as the farm buildings help to shelter the house from the west. The fields at the bottom of the slope, bordering Breazle Water, a tributary of the Tamar, provide good shelter for cattle. Water from all the ditches and drains flows eventually into Breazle Water. The soil is heavy loam, the sub-soil in some parts clay.

The first known reference to Chimsworthy, as *Chemesworthi,* is in an Assize Roll of 1286. The name means 'Ceolmaer's settlement'. It is derived from an Anglo-Saxon personal name, 'Cemmi', a pet name for Ceolmaer which comes from Ceorl, meaning a rustic. 'Worthy' means an enclosure, hence 'Cemmi's enclosure' or 'Ceolmaer's settlement,' and the farm dates back in all probability therefore to some time before the Norman Conquest.

The house lies close to the farm buildings, the original shippon being continuous with the house. A large barn forms the second side of a triangle, while four sheds form the third side between the house and barn, and a 5-ft. wide arch between two of the sheds makes a way in and out of the yard at the bottom. At the top end is the driveway and bank with the old well in it and beech hedge above.

Outside the old shippon which now houses the Jersey bull, is a lean-to shed where

37

the trap used to be kept. Below this is a big flagstone and built over this was a bucket-type lavatory. This was later replaced by the flush one on the yard side soon after water was piped from the later well during the last war. It is still in use though the water supply is now from the general borehole one.

FIG. 2. Chimsworthy, Bratton Clovelly (Devon)

Chimsworthy is mentioned again in 1376 and in 1533–38 in a lawsuit in the Court of Chancery. On a map in the Devon county archives it appears as Chelmsworthy as it does also on the ordnance map of 1809. However, on the Land Tax Assessments from 1794 to 1808 it appears as Chimsworthy. In these assessments the tax was 12s. a year up to 1789 when it was increased to £3 4s. 0d.

In *Devon*, W. G. Hoskins describes Chimsworthy as 'mainly an Elizabethan farmhouse which has been turned around so that the original front door is at the back. It is possible that the core of the house is medieval.'

A WEST DEVON FARM

Under the Town and Country Planning Act of 1947 Chimsworthy is included in the schedule of buildings of Historic and Architectural interest as a Grade II monument, and is thus described: 'Plastered rubble with slate roofs and plastered stone stacks. T-shaped plan with two storeys. Wood casements and two granite mullioned windows. Plain massive collar-braced roof. Open fireplace in kitchen. Sixteenth century and later.' It is still on the list under the Act of 1962. The north end of the house, that is the horizontal stroke of the T, is the oldest part, and the vertical stroke represents the later Elizabethan part.

This account would not be complete without mention of the book *The Red Spider* written by S. Baring Gould and first published in 1887. It is a story depicting the daily life of the parish of Bratton Clovelly where the author spent his early years. The hero of the story lived at Chimsworthy which figures prominently. As Baring Gould was born in 1834 it is reasonable to suppose that the descriptions refer to the farm as it was in 1840–45.

One of the most interesting things about the house is the four oak 'principals' in the later part of the house. These are each made from two continuous oak beams reaching from the apex in the roof, to the ground. These can be seen upstairs but downstairs, although the outlines are there, they have mostly been plastered over except in the old shippon.

In the year 1900 the roof was raised by about 2 ft. The line along the wall can be clearly seen on the west side. The existing thatch was replaced by slates. In the next village lives a man who worked on this job when he was fourteen. When removed, the thatch was found to be 4 ft. thick and woven in and out like a basket. During about 60 years of work in old houses, this man says that he has never seen beams like those in Chimsworthy. About these Baring Gould says: 'Adjoining the house was a good oak wood covering the slope to the brook that flowed in the bottom. Fine sticks of timber had been cut thence time out of mind. The rafters of the old house, the beams of the cattle-sheds, the posts of the gates, the very rails, the flooring, all were of oak, hard as iron; and all came out of Chimsworthy wood.'

Since 1959 there have been many changes. Electricity has been installed and a bathroom and lavatory made out of one of the bedrooms. The open fireplace in what was the old kitchen has been modernized and made more efficient, though it has been done so that the original brickwork is still intact. This is now the main living-room. The old granary was converted in 1965 to make an extra bedroom for a worker, and a more useful room has been made from the old dairy.

Except for ditching, for Chimsworthy is inclined to be wet, there have not been very great changes as far as the fields and hedgerows are concerned, though scrubland has been reclaimed and put in grass and in two instances one field has been made where previously there were two separate ones.

The barn is the same as it must have been for years with the addition of two extra doors. At the time of the village revel and more recently in harvest time, dancing took place in it.

In the early 1920s the present shippon was built below the barn. It is now rather a ramshackle building of concrete blocks and galavanized iron sheets. The thatch on the barn and stable roofs has been replaced by sheets also. The stable is now an extra shippon for 11 cows. A milking parlour was built in 1959.

A large three-bay barn was built in 1966 in part of the orchard near the Germans-week road entrance. This entrance is basically as Baring Gould knew it . . . 'An avenue of contorted, stunted limes led to the entrance gates of granite, topped with stone balls . . .' At the other entrance there are also granite posts but the stone balls have gone.

There are many springs on Chimsworthy. The water runs almost perpetually in several of the ditches, providing drink for the cattle and flowing ever down the slope to the brook in the valley at the bottom.

This then is Chimsworthy in 1969. Generations of farmers and their families have lived here, walked up and down the cobbled yard and passed the granite window on the stairs on their way to bed. They have watched the changing lights on Dartmoor and seen the sunset over the Cornish Tors. Thousands of animals have grazed in the fields and walked in and out of the yard. For perhaps a thousand years there has been a farming unit here. By modern standards the house cannot be called comfortable or warm, but it is a tribute to those men of centuries ago who hewed out the massive oak beams. Yet illogically, the fact that it still stands four-square to the wind and Devon rain when so many who knew it are gone, illustrates the comparative transience of man.

A Dartmoor Farm

Lake Farm lies well into Dartmoor, though close to the Ashburton–Tavistock main road. Most of the roads on Dartmoor are, however, of comparatively recent origin, and farms that today seem reasonably accessible were only a hundred years ago remote and rarely visited.

The present farmhouse is a very fine example of the ancient type of 'long-house'. Dartmoor still retains many good examples of this type of farmstead which was once so common in the upland parts of Britain. In these houses the family and the cattle were housed under one continuous roof, with the dwelling-house at the upper end and the cattle-end (called the shippon in the West Country) at the lower end for obvious reasons of drainage. In the oldest of these houses there was a common entrance for both humans and cattle, wide enough to enable the cattle to enter a 'lobby' and to turn into their stalls. In later examples, like Lake, an internal wall was built between the house and the cattle, and separate entrances constructed. Usually, however, there was a doorway in the internal wall which gave direct access to the cattle from the house without going outside.

This type of house, of immemorial antiquity, was especially suitable in upland areas where the winter climate was tough. It conferred benefits on both man and beast to live under the same roof, not least because the cattle could be sheltered and tended at particularly difficult times, and because the cattle in turn helped to keep the house warm.

The earliest reference to Lake as a farm occurs in 1332, when the owner William Atte Lake is named in a tax assessment. It had probably existed for some time before this, but is not as old as many West Country farms which go back in thousands of instances to Saxon times. Domesday Book (1086) shows that the manor of Spitchwick in which Lake lay had only the lord's farm and eight peasant farms at that date, and I do not think that Lake was one of these Saxon farms. Most of the manor was still under dense woodland or upland moor and was probably kept mainly for royal hunting. It belonged to Earl Harold in 1066, and after his death at Hastings it passed to William the Conqueror in person.

Later, under the pressure of population in the thirteenth century, more farms were cleared and settled, and it is at this stage that farms like Lake and Aish, which Mrs. Wilkinson talks about in this essay, came into existence. Many of these medieval farms were small and poor. Even in 1848 (tithe award) Lake was only 46½ acres, and the two Aish farms (Higher and Lower) were only 60 acres between them.

The Hamlyns, who lived at Lake for so long, took their name beyond much doubt from the Hamelin who held more than a score of manors in Devon and Cornwall after the Norman Conquest. A junior branch of this family appear at South Tawton, on the edge of the moor, well before the end of the twelfth century, and then proliferated into numerous branches on a lower social level whom we find on many farms on and round Dartmoor within the next 100 years.

It is not uncommon in England to find farmers bearing the names of Norman and later great landowning families, and indeed labourers also, for the simple reason that only the eldest son inherited the ancestral estates in most parts of England, and younger sons had to make their own way. As the generations passed their descendants could naturally be found at all social levels. There are still Hamlyns living in the parish of Widecombe-in-the-Moor and other parishes round the moor, and others scattered all over Devon, according to the current telephone directory. But all are descended from that adventurer who came over with William the Conqueror to chance his arm in the Great Plunder of the Norman Conquest.

*

LAKE FARM, POUNDSGATE
by Freda Wilkinson

Lake Farm (or rather Lake and Higher Aish farms, for both are now part of the same holding) lies just north of the hamlet of Poundsgate in the Manor of Spitchwick, in the parish of Widecombe-in-the-Moor. The two lowest fields lie on the 700-ft. contour, on the east side of the road through Poundsgate, the rest lie in a block extending from the west hedge of the same road to the open common, and Dr. Blackall's Drive in the west (about 900 ft.), and are bounded by Lower Aish land to the south and Lower and Higher Tor lands on the north.

Most of the original Lake Farm lies on the granite, but the Higher Aish fields lie on what we call the blue elvan. This has a significance to the stock farmer as the pine or 'moorsick' caused by cobalt deficiency in ruminant animals, particularly sheep, occurs in the granite area but not on the elvan. The surface granite 'moorstone' is also a much easier material with which to build.

The farm is now just over 68 acres, now all in permanent grass but scheduled as about 43 acres arable, 13 acres rough pasture and 12 acres pasture—of the latter about 6 acres are meadow. The fields are small (like all those in the parish that have never been occupied by gentlemen farmers), averaging about 2½–3 acres; the largest are just under 5 acres. There are common rights on Spitchwick manor for which, until the 1920s a 'chief-rent' of 4/5d. was paid annually to the lord of the manor.

Lake Farm was occupied almost continuously by the Hamlyn family from 1545 till
42

1958; they also farmed Higher Aish and Lower Aish for very long periods. Lake, Higher Aish, and part of Lower Aish are now owned and occupied by members of the Wilkinson family.

The Hamlyns bought Lake in 1545, though they had other farms on Dartmoor for long before that. Their name appears in Dartmoor records as early as the twelfth century. They were forced to sell most of their farms, including Lake and Higher Aish, in 1745— we do not know why, except that the records show a growing tale of mortgages for

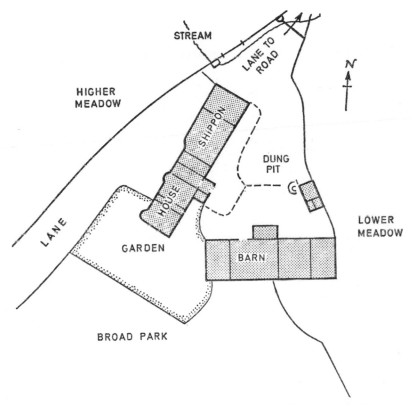

FIG. 3. Lake Farm, Poundsgate (Devon): plan of site

several years before that. They came back as tenants at intervals after this sell-up, until the last of them died at Lake within recent years. Lake was sold as part of a larger estate in 1958 and came into the possession of the Wilkinsons, who still farm it.

The plans show the farmstead as it was in the 1930s. The little back kitchen or store-room next to the entrance porch had a piped cold water supply brought into it between 1921 and 1958; before that water for the house was dipped from the trough near the north end of the shippon. There were originally two doorways to this room; the one from the porch was blocked up (leaving a window) in 1959, and the one into the big

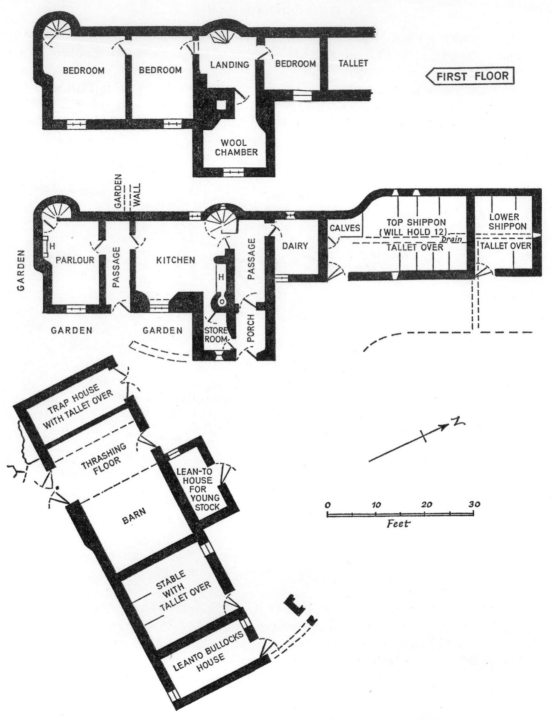

FIG. 4. Lake Farm, Poundsgate: plan of house and buildings

kitchen opened up at the same time. The latter had previously been blocked up since before living memory. Otherwise the farmstead as in the plans is probably much as it was in the time of the Thomas Hamlyn who had the porch, and probably the whole house, built and inscribed with his initials in 1661. Although the shippon adjoins the house there is no connecting inner doorway, nor sign of a blocked-up one, nor is there a back doorway opposite either of the outer doors of the farmhouse; it was presumably built after the true 'long-house' era.

The two rooms (parlour and the bedroom over) at the south end, could have afforded a self-contained unit approached by the passage from the southern outer door, as there is a winding stair from the lower to the upper room, in addition to the similar stairway from the kitchen to the upper floor. Thus the house could have contained two families. Perhaps the smaller part was occupied by the retired grandparents of the farm family.*

All the walls are thick, being of double-faced granite construction, mortared, with loose rubble and small stone between; but the east wall of the kitchen, on each side of the window, is twice as thick, rather curiously, as it only appears to have the extra space between the facings filled with small stone. There is also great apparent thickness in the walls at the west end of the woolchamber, but this is accounted for on one side by the chimney of the great hearth below. There is a bread oven built inside the east end of the hearth.

The beams, of rough-hewn oak, are exposed in the ceilings of all the rooms; the bedrooms, parlour, and dairy are plaster-ceiled between the beams, and, in the upper floor, below the narrow loft space. In the kitchen the plastering of the ceiling extends about 2 ft. 6 in. from the south end; the rest of the ceiling is the exposed floorboards of the room above. These were renewed in 1959, when the old oaken planking, about 18 in. wide, was taken out as it was rotting. That may have been there since the house was first built. The height to the beams in the lower rooms is just over 6 ft., the door lintels are just under 6 ft.

The floors of the lower rooms have been cemented over at some time, but the porch floor is of unsquared flagstones, cemented between. It is said that within living memory the southern interior wall of the kitchen was lined with carved panelling, or possibly a carved screen, originally taken from the old chapel of St. Leonards at Spitchwick and that it was removed many years ago by a previous landlord either for his own house or for sale. The winding stairways are of granite, the one from the kitchen having wooden treads and risers over the stone.

The house, with its modest adaptations to twentieth-century living, is comfortable and comely. The thatch roof, though expensive in upkeep, keeps it well insulated, and the walls, though they have no damp course, are surprisingly dry.

Those who have never actually lived with, and cooked on, an open hearth, might prefer to see the kitchen hearth unencumbered with a Rayburn stove, but for the time

---

* This is undoubtedly the purpose of these end-rooms. I have seen similar arrangements in other isolated farmhouses (W. G. H.).

it awaits the day when someone manages to invent a method of restraining an open-hearth fire from blackening the whole kitchen with smoke and persuades it to do the cooking and heat the bath water without a great many minions to serve it.

The higher and lower shippons (the whole farmstead site slopes slightly down towards the north-east), are divided by a granite partition wall as far as the tallet (loft) floor; the tallet covers both, and has a step down in its board flooring, to the part above the lower shippon. The galvanized roof of the shippon is a little lower than the thatch of the house. The tallet is reached by a ladder from the higher shippon and has a door in the north gable end for hay, etc., to be unloaded into it. There are rack holes in the tallet floor, against the walls, for hay to be thrown down to the cattle feeding racks. These have wooden covers.

The higher shippon has its top corner boarded off to make a calves' house. It has a floor of rough granite flags with a shallow open drain 3 ft. wide, down the middle, which narrows and goes through the wall into the lower shippon. There are standings for 12 cattle with wooden rails between the pairs. About 2 ft. out from the walls a granite curb, or kerb, makes an edge to the front of the standings. The tying posts stand in holes hollowed into the stones of this curb, and their top ends are pushed through holes in a timber near the ceiling and fixed down with a wooden pin or a nail. Thus they can be removed or renewed by taking out the pin, sliding them up far enough to be taken out of the curb holes, and then pulled down out of the top holes.

The beams supporting the tallet in this shippon are only just over 5 ft. above the floor. The whole farmstead is rather hazardous to anyone above average height.

The lower shippon has its floor (which is of flagstone and cobbles) on a slightly lower level, leaving a little more headroom. This has standings for 3 bullocks, rather longer than those in the higher shippon. It may have been designed for the working oxen, a fully-grown steer being generally larger than a cow; there would have been room behind them for yokes and other gear to be hung up. The gutter here, after receiving tribute from the higher shippon, continues under the doorstep and under the cobbled way outside, to the dung pit beyond. Both shippons have half doors and slit windows, and the higher shippon door has a cat's-hole in its lower half.

The barn has a boarded threshing floor, raised on sleeper beams between the two doorways. The south-facing double doors leading to Broad Park are wide enough and high enough to allow a wain or a tractor trailer to be backed in on to the threshing floor so that the load could be thrown off on to mows on either side. These doors close on to a centre post which can be removed. When the barn was built wheeled vehicles were not used in this district, but a pack-horse with crooks loaded high and wide with corn would need practically as large a space to pass through as would a loaded vehicle.

All the buildings are very ill-ventilated by modern standards but the old people thought that the place for fresh air was out-of-doors and that cattle should be kept as snug as possible when they were lying in.

The last Mr. Thomas Hamlyn of Lake was very proud of his horses and he used to stack manure inside the stable, behind the horses, to warm the place up and make their coats shine. No doubt it did that, but the reek and fug wouldn't be thought at all healthy nowadays. One or both of the little fuel-houses were once pigs' houses. One still contains a granite trough and just outside them, in the yard, stands a huge round boulder which has been hollowed out at the top to make a sort of Cyclopean pigs' trough where the dairy wash and kitchen waste could be thrown. Outside the porch are two half-hundredweight granite stones, with rings in their tops, which were used to weigh the wool against.

The yard is cobbled in front of the house and buildings. This cobbled area ends with a curb which encloses the dung pit. This was a very important place in the farm economy. Good farmers, in the days when thrift was understood and artificial manures unobtainable, composted, with the dung, any vegetable material which couldn't be eaten and which would rot down, for use on their land. Worked-out stubble and beat often provided a basis for the heap. Way-soil was collected from the parish roads. Vancouver, writing in 1808, noticed that potato-haulm was often spread on the roads to be pulverized and mixed with manure by passing traffic and then collected again for the dung heap.

There are three meadows on Lake. The old meaning of 'meadow' in these parts at least, was grass land which could be irrigated, usually by artificial gutters, during the winter, to provide early grass, mainly by keeping the temperature of the soil above freezing point. This gave an early bite for the ewes and lambs in late March and early April, and was sometimes used for the milking cows as well, though they were more apt to 'paunch' (poach) the ground until it had time to harden, and to tread in the gutters. These few weeks have always been the hungriest time on and around Dartmoor, when fodder is getting short and the fields are still winter bare. Suitable meadows were usually, but not necessarily, laid up in mid-April to be mown later for hay. The practice of watering meadows is a very old one in Devon. In most parts of the country it appears to have started in the eighteenth century, but Colepresse (1667) writes of it as usual and well-established practice in Devon. 'About Allhallowtide those that have (or can make) the convenience of water, doth water their meadows with such water principallie as passes nere any dung heaps, or mixens, staules, slimie ponds, free and mirey highways, or hot springs till the middle of May: eastward but till Candlemas.'*

Similar artificial gutters fulfilled an important function by supplying 'pot' water for the household and livestock at Lake, at the Post Office (which was previously the village bakehouse and before that the blacksmith's shop), and at Lower Aish farm. All now share with the rest of the village a council water supply which has its source in a spring in Tongue meadow on Lake. This is now a rushy rough pasture but was a watered meadow before the small spring there was tapped by the Council.

An attempt was made to work out the age of the hedges and thus the approximate

* This artificial watering of meadows was well-established practice in Devon in the time of Elizabeth I (W. G. H.).

date of the original enclosure of each field by counting the species of trees and shrubs in each hedge. This, according to the theory worked out by Drs. Hooper and Hoskins should give roughly 100 years of age for each species. The numbers varied between five and nine, or, if bramble, briar and furze were included, between seven and twelve. The tree species consisted of oak, ash, whitethorn, blackthorn, nuthalse (hazel), holly, sycamore, elder, quickbeam (rowan), birch, and a few beech.

However, dating by this method did not appear to agree very well with other possible methods. For example, one would imagine that the hedge between Higher and Lower Hill parks would contain considerably fewer species than the hedges surrounding the two, but this is not the case. One would have expected that the number of species would remain constant along the continuous hedge to the south of the Whiddon Downs but it does not.

Perhaps the folk who first made these hedges planted whatever species were available and liked variety. The hedgerow theory would probably be useful even here if it combined with other data. For example, the size of stones in the banks—a large proportion of massive stones would no doubt have been shifted during the original clearing and incorporated into the first hedges, whereas later subdivisions would have been largely faced with smaller stones turned up by subsequent ploughing. The relative depth of lynchetting by soil-slip over years of cultivation might also give an indication of age in hedges following the contour, though this by itself might mislead as sometimes a day's torrential rain on a bare tillage field might wash as much soil to the bottom hedge as would have gathered there throughout centuries of careful husbandry.

3. Chimsworthy, Bratton Clovelly

4. LAKE FARM, POUNDSGATE

Two Dorset Farms

A North Dorset Farm

The history of Liberty Farm is tantalizing. It changed its identity towards the middle of the eighteenth century, not only its name but the names of all its fields also. The result is, as Mrs. Lemmey shows, a fascinating bit of social history (and a confusing one for place-name scholars) but it only enables us to get back with certainty for rather more than 200 years. She has not yet identified the farm under its former name in a survey of 1684. Once this has been accomplished—and I think it will be solved one day —we shall be able to trace its history backwards for many more centuries.

Even as it is, we know that the land of Halstock—which means 'the holy place'— was already settled and being farmed when the king of the West Saxons granted a large estate here to a subject, as long ago as the year 841 or thereabouts. We know the boundaries of this estate. They have been worked out and are practically identical with the boundaries of the modern parish (now 3,181 acres). So what is now Liberty Farm formed part of this estate, which shortly afterwards was granted to Sherborne Abbey not far away. Sherborne held it until 1539, and at the Great Plunder that followed the dissolution of the monasteries it came into the hands of lay landowners. The change-over of land ownership at the Dissolution was the greatest act of plunder since the Norman Conquest.

But there is another fascinating possibility. About half a mile south-west of Halstock church, and a mile and a half south-west from Liberty Farm, the remains of a substantial Roman building have been unearthed. This was either a Roman villa or a large farmstead of the same date, surrounded by a more or less considerable estate. On Abbot's Hill Farm, between Liberty Farm and the villa-site, is a field still called *Stanchester*, which means 'stone fort'—a Roman site about which nothing else is known. But it does suggest that the cleared land—the estate of the villa—extended well towards the present Liberty Farm, if not indeed right over it to the natural boundary of the stream that forms the eastern boundary of the farm.

It is quite possible that this estate, already cleared and settled, was taken over *en bloc* by some Saxon newcomer—after all it was the natural thing to do, to take over well-farmed land rather than a wilderness of trees and undergrowth—and that is perhaps why it appears as a going concern in the ninth century. This was a fertile and well-watered tract of land, attractive to early farmers, and there was no reason why it should ever have gone out of cultivation

We cannot be certain of these things yet, but it seems to me highly likely that in many favoured parts of England there has been an unbroken continuity of farming since

Roman times or even earlier; and I also think that Dorset, for various reasons, is one of the likeliest parts of England for this continuity. If so, Liberty Farm, despite its recent name, may have been farmed continuously since Roman times. But at the moment this can only be an hypothesis, though a very promising one.

*

LIBERTY FARM, HALSTOCK
by Pam Lemmey

Liberty Farm lies in west Dorset $1\frac{1}{2}$ miles from the village of Halstock and $1\frac{1}{2}$ miles from the main Yeovil–Dorchester road. The farmstead stands 300 ft. above sea-level, half-way down a hillside, sheltered from the south-westerly winds, the land running down to a stream in the valley. The water supply gave no problem as there are springs in several fields and one a few yards from the house. In addition small streams run down the 'goyles' to the valley, providing further watering places for stock.

The soil is medium loam overlying yellow clay. The eastern boundary of the farm is the stream (a tributary of the Yeo) in the valley. The western boundary is formed by a lane which runs along the ridge of the hill. It is possible to trace a paved track, now overgrown, up from the village of Halstock, which meets this lane opposite one of our farm gates, and it seems likely that this track continued down the hedge, through the yard, down to a ford across the stream and thence to the Dorchester road. A footpath on the O.S. map follows this route. This would have given the farm more direct access both to the village and main road than it has now.

To explain the field names and the name of the farm itself, it is necessary to refer to its most interesting owner, Thomas Hollis, who acquired this land and several more farms in Halstock and Corscombe in 1740. Hollis, philanthropist, scholar, radical and amiable eccentric, was a keen upholder of liberty, and detested corruption and tyranny in all its forms. He made many benefactions to universities in Europe, and after 1758 gave large sums yearly to Harvard in America. He was responsible for the republication of the works of many of those authors whom he felt shared his ideals, and he re-christened the farms which he purchased with the names of these authors. We thus have in Halstock farms named Locke, Sydney, Marvell, and Neville. Another he named Harvard, and it seems in character that he gave this farm the name of Liberty, although Fägersten in *Dorset Place Names* suggests that it may be connected with the fact that Halstock was a Liberty. But it is not traceable by this name prior to Hollis's ownership, and it seems likely that he bestowed it on a farm with an older name.

He also named the fields of his farms by the names of 'heroes and patriots for whom he hod veneration'. Liberty's field names are Cassius, Messala, Confucius, Lysurgus, Socrates and Maber (this last being the name of his favourite steward). On some of his

52

other farms he named the fields in more abstract fashion—Toleration, Government, Constitution, Commonwealth, Revolution and so on.

This has, however, made the tracing of the farm's earlier history extremely difficult. There exists a survey, made in 1684, of all the property in Halstock, which then belonged to the Fermor family. It gives much useful detail regarding tenants, field names, etc. Unfortunately it has proved so far impossible to identify Liberty Farm from the information there given. Two or three farms are recognizable, and I hope that I shall identify in time the farm I am now writing about.

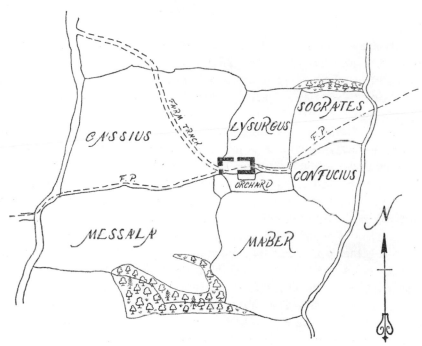

FIG. 5. Liberty Farm, Halstock (Dorset)

This farm, together with the rest of the manor of Halstock, was granted to Richard Fermor after the Dissolution, before which the lands were held from an early date by the abbey at Sherborne. Though Halstock is not mentioned by name in the Domesday Book, there is in fact a charter, dated 841, whereby Ethelwulf, King of the West Saxons, grants lands at *Halgan Stoc* to the deacon Eadberht. This is of interest as Grundy suggests that the 'Pigstye' brook which marks the eastern boundary of the charter is the same stream which forms the boundary of the farm. The bounds of the charter then go west following the farm boundary (as it was before 1920) until the lane is reached. Another charter dated 998 shows that Sherborne Abbey had acquired Halstock, which they then held until 1539.

At this point it should be mentioned that a brief shrub count in the internal hedges of

53

the farm shows several with 11 different species, one with 12 and one with 14, in a 50–100-yd. sample, showing a settlement here of some considerable antiquity. Indeed a count of 14 species would suggest that some hedgebanks may go back to the sixth century.

After the death of Hollis in 1774, the estate was bequeathed to his friend Thomas Brand, who thereupon took the name of Hollis. In 1804 the lands were willed to the Rev. J. Disney, who in 1820 sold part of the estate including Liberty Farm to the Earl of Ilchester, in whose possession it remained until 1920 when the lands in Halstock were sold, and Liberty was split into several smaller holdings. The farm now consists of 67 acres.

A beautifully drawn map, dated 1799, shows the acreage of Liberty Farm at this date as 313 acres. This included 105 acres of Buchanan Farm which was occupied by the same tenant, and of which there is no trace today, nor is it shown on the tithe map, which again shows Liberty as 313 acres. On both maps most of the land is pasture and meadow, a few fields only arable. The field boundaries on each map are identical, and indeed it is possible to trace exactly the same field lay-out in an aerial photograph taken in 1947. Several gateways on the early map are still in existence, but neither map shows the present track down to the farm. This must have been a later addition.

The farm buildings and house are grouped around a yard, but with access from several directions, as would be expected with a stock farm. The buildings include a fine old thatched barn with hammer-beamed roof, a cowshed, and until recently another large thatched barn close to the house. These buildings have now been adapted and extended, and together with a new parlour and silage barn provide accommodation for 60 head of dairy cows and followers and their winter keep.

The house, which is L-shaped, is thatched and built of local stone, reputedly quarried from Liberty Goyle nearby. The West Dorset volume of the Historic Monuments Commission lists the farmhouse as of seventeenth-century date, with eighteenth-century additions. It is of the long-house type, with a cross passage. The two rooms at the west end of the house seem originally to have been used as outbuildings, and are indicated as such on the tithe map. Certainly the end room was used during this century as a cider cellar. This end of the house has also been used as a separate cottage, and the downstairs portion now comprises my mother's flat.

The room on the west side of the passage was presumably the parlour as the floor is boarded. This is now the dining-room/office. The room on the other side of the passage, with flagged floor, open fireplace and baker's oven would have been the kitchen. This is now our living-room. The other rooms on the short side of the L must have been service rooms. These are now kitchen and pantry. Until last year, when alterations were made, the pantry was double the size of the kitchen (this has now been reversed) and this pantry was recently used as a cheese-making room, the cheeses being stored in the room above, a trap door allowing access. The present back kitchen appears to be original, but the dairy and back porch are a later addition. During alterations in the back kitchen we discovered on removing an old stone sink, that beneath it lay a hole about 6 ft. square and 5 ft. deep with bricked sides and ceiling. It had an earth floor, and a small arch about 2 ft. high and 1 ft. wide at ground level. This was filled behind. The use of this hole is a mystery to us. It could perhaps be connected with drainage.*

One of the most attractive features of the house are the Ham-stone mullions, of four lights with ovolo mouldings—these are original. The upstairs windows are 'eyebrow' dormers.

The first mention of the actual dwellers in the farmhouse is found in the Halstock Parish Register, when the burial of 'Thomas Guppy of Liberty Farm' is recorded in 1780. From the map of 1799 we find that the then tenant was Joseph Guppy. These Guppys were the most prolific family in eighteenth-century Halstock and it is difficult to disentangle the various branches. However, Joseph's marriage to Joan Rogers in 1790 is recorded, as is the birth of a son, Thomas, in 1796. Another entry in the register tells of the death of 'Mrs. Joan Guppy of Liberty Farm', but there is no further trace of Joseph or Thomas.

The churchwarden's accounts show a Thomas Miller at Liberty in 1815, but by 1829 the tenancy had passed to William Dodge, in whose family it remained until c. 1887. William, who had at least six surviving children, died in 1845 aged 80, and the farm was then run by his sons George and Simon, their sister Elizabeth living with them. None of these three married and all lived to a ripe old age, dying at 79, 81 and 95 respectively. The 1851 census shows Simon as the head of the family (he was the eldest of the three), but in the 1861 census George is shown as the head and there is a note to the effect that Simon was deaf, perhaps the reason for his losing the ascendancy.

After c. 1887 the farm passed to Dan Wrixon and early in this century to the Biggins family. Then followed the Dredges and Edwards until 1962 when the farm was bought by us.

This account seems barely to have scratched the surface of this farm's history. There is so much more to learn, particularly of its early history; but even now one can say it has a documentary history going back to the ninth century. Some hedgebanks may be even older, as the charter of 841 already indicates settled land, and the evidence of the shrubs and trees seems to bear this out.

---

* It is almost certainly an old cesspit and drain, possibly belonging to an earlier house (W. G. H.).

A Farm in Purbeck

Godlingston Manor is a good example, among other things, of how difficult it sometimes is to get material for the early history of a place. It is not mentioned anywhere by name until as late as 1299, and even that record (an inquisition post mortem) tells us the very minimum about it. Yet, by knowing how, one can trace its history back for certain to just before the Norman Conquest.

It is not mentioned separately in that great survey, Domesday Book, which was completed in 1086; but it is virtually certain that Godlingston was included with the entry relating to the manor of Swanage. It then belonged to the wife of Hugh fitz Grip, who held most of the land around the Swanage district. A little later we find her succeeded by a family called De Lincoln, and in a record of 1166 we read that Alured Talbot held one knight's fee under Alured de Lincoln. As the Talbots appear in later Godlingston records down to 1367, we can safely infer that Alured Talbot was the owner (mesne tenant) of Godlingston within a hundred years of the Norman Conquest. With a gap, therefore, since we do not know exactly when the De Lincolns took over from the fitz Grips, we can take the history of the farm back to 1086, and then back to Saxon days before the Conquest when an Old English landowner—Alward—had the estate.

Godlingston has had nine owners between 1066 and the present day, not counting Alward (as we know nothing of his family)—in other words, an average of one owner every 100 years. One wonders how this compares with other English estates. There are, of course, a few authenticated examples of estates which have never changed hands since the Norman Conquest—for example, the Kellys and the Fulfords in Devon still own and live on the estates they are credited with in Domesday Book. So do the Shirleys of Ettington in Warwickshire. Mostly, though, estates have changed hands several times since the Norman Plunder, more rapidly in some centuries than others.

At Godlingston the Talbots stayed for at least 200 years. They were already there in 1166, and they sold it in 1367. Then came a number of changes of ownership—not always entirely clear—until the Wells bought it in 1557. Since then—over 400 years—there have been only three owners: Wells 1557–1687, Frampton 1687–1765, and Bankes from 1765 to the present day. Tenants, of course, change more rapidly, though there have been examples in the past of tenant-families who have farmed the same farm for 200 to 300 years. One would like to know more about this side of English farming history.

To come back to Godlingston, however, we can trace its history even farther back than the Saxon Alward. The name means 'Godelen's farm' or, if it is a feminine name, 'Godelena's farm' though even he or she may not have been the first owner. The original

56

farm or estate (as it was a small manor on its own) was once much bigger but was considerably reduced in size within living memory, as Mrs. Bowerman shows. It included every sort of soil—chalk downland, rough grazing on the heath, arable on the Wealden clay, and woodland and meadows by the streams. The house was well sited—facing south with the shelter of the great hill behind it, and on the greensand where springs supply never-failing water.

Godlingston Hill, behind the house, rises to 655 feet, and commands a tremendous view, ranging from Wareham to the north right around to Portland Bill far to the south-west. There are many prehistoric barrows on the hill and on the Godlingston Heath to the north; and on the southern slopes of the hill there survives a system of strip-lynchets, i.e. old cultivation terraces. Farming has gone on in these parts since Iron Age times, and perhaps even earlier.

The fortified manor house must have been rebuilt by the Talbots, if as Mrs. Bowerman says, the massive tower at the west end of the present house may be dated as c. 1290.

Tenants change more frequently than owners. During the 200-year ownership of the Bankes family there seem to have been ten tenants—an average of one every 20 years. The Bowermans came in 1949, so they have reached the average at least. When they took over, the farm and buildings were neglected—the Army, as usual (Dorset has suffered heavily from the Army and still does)—and most of the land had not been ploughed in living memory. Gorse and bracken covered a great deal of it.

The Bowermans decided to farm Godlingston as though they owned it and have now brought it up to modern standards of husbandry. Part of the farm had been occupied by the Army during World War I, but when they departed in 1918 they left behind excellent wells and a good water-system. This has been extended to give an adequate water supply to every field. During World War II the Army had 200 acres of the hill-land as a training area.

The present farm is still large by English standards—519 acres*—but the original Saxon and Norman estate was something like 2,000 acres. Unfortunately the field-names are the usual rather dull lot—names like Ash Close, Cowleaze, and so on. They must have been changed several times over during the long life of the farm. One would like to know what the medieval and the Saxon fields were called, but these are lost to us for ever.

*

GODLINGSTON MANOR
by Jean Bowerman

Godlingston lies about a mile north-west of Swanage. It is situated, like several other underhill farms of Purbeck, where the greensand belt separates the chalk from the

* Now (1969) 480 acres.

Wealden clay. Here is the source of a spring which flows southwards towards the house. The site is well sheltered under the chalk ridge to the north, and faces the sun.

The finding of Stone Age implements, Romano–British fragments, and the presence of ancient tumuli on the hill show that people have been farming around here for a very long time.

The farm was formerly a manor which stretched from the lane south of the house, to Poole Harbour to the north, a distance of 2½–3 miles. The width of the manor just included Brands Bay, on Poole Harbour, so giving access to salt water, and a small stream formed the eastern boundary. There is a kink in the line of the land where it passes over the North Purbeck Downs. Here the boundary stones are marked on the 1839 tithe map for the Parish of Swanage. Basically the field boundaries have hardly changed since 1775, but the extent of the present farm is only about half of the original manor. The division came with the making of a new farm on the North Downs between the Great Wars, but the east and west boundaries being parish bounds, have not changed at all.

The 1839 map shows various scattered parcels of land which are now no longer part of the farm.

The site chosen for the original manor had a normal layout for the area. Heath on the Bagshot beds, downs on the chalk, arable on the Wealden clay, woods and meadows by the streams.

The western boundary is that of the east field of the neighbouring Knitson open field system, whilst the south boundary is marked by an ancient water lane running underhill from Swanage to Corfe.

Along this southern boundary hedge are examples of nine different kinds of trees, which together with the high banks appear to indicate Saxon origin: ash, maple, blackthorn, oak, elm, holly, hazel, whitethorn, dogwood.

The house is situated almost on the southern extremity of the manor where there is water. About the year 1890 the farm buildings were ranged to the west and east of the house, but those to the east were destroyed by fire. Subsequently more modern buildings have been erected to the west of the house.

An authority on thirteenth-century domestic architecture, Miss Margaret Wood, dated the house at 1290. It was at one time fortified, and a semi-circular tower remains. In the back garden we have unearthed a medieval water sluice, and a square of stone paving no bigger than a yard (36 in.) with remains of undressed kerb round it. We are awaiting an expert's opinion on remains of a large stone archway with a door recess.

If our theory proves to be correct, that the latter is part of the original entrance, then this archway could very well have been set between two towers, with a loop light (which we have unblocked) commanding its north side. The remaining tower has all the aspects of a gatehouse structure with its arrowslit* apertures on one side only. These with a corresponding set in a second tower would guard the entrance way.

* In their present form these are too wide for arrows, more like musketry openings. Regarding a

A FARM IN PURBECK

The house has evolved from a medieval hall, an upper storey and dormer windows being added in the sixteenth or seventeenth centuries. Inner brick walls have now divided it into several rooms and passages. A kitchen wing was added in the eighteenth century, when the roof was also rebuilt.

We are still hoping to pin-point the dates of the latest alterations and rebuilding done in the last century. The joists (except in the kitchen) are pine, probably from the time of the Baltic Wars when oak was difficult to obtain. The doors are extremely disappointing both in design and material.

In 1861 there was still an arch similar to the front door archway and placed immediately opposite it inside the house, and 'a very rude stone stair which conducts to the upper floor'.

In 1773, Hutchins, the Dorset historian wrote, 'Here was anciently a chapel which now makes part of the house'. By 1861 it had apparently completely disappeared.

There is considerable evidence to suggest that there was an ancient moat around the house. An estate map of 1775 shows a watercourse flowing just to the north of the house and feeding two stretches of water on the east and south. The tithe map of 1839 shows a long pond on the west of the house.

In the 10-acre wood behind and above the house to the north, the springs in the greensand have been diverted from their natural course by several long stretches of man-made ditches, at an unknown date, which lead the water to the back of the house. This diversion of water could have been for both domestic and defensive use.

Today the watercourse flows down the natural lie of the land from the source, which means it completely misses the house and continues in a straight line to the main ditch, which carries it on to Swanage.

The original source of water comes from deep water springs around the base of a sandhill in the north of the wood. The geological strata have a 70-deg. dip in this part of Purbeck, so the springs are from very deep sources indeed. We prefer this water for drinking, as apart from the flavour it is always delightfully cool.

second tower, we are also considering the possibility that it could have been sited on the north side with a connecting passage from present undated high opening.

Hampshire and Gloucestershire Farms

An Old Hampshire Farmhouse

Mrs. Beresford's account of Hatchetts Farm concentrates on the evolution of the farmhouse to its present form, from about 1500 to the present day. Her history is valuable because she brings out the fact that an English farmhouse often has a complicated building history, sometimes as complicated as that of the parish church. Often it is difficult to work out the precise history of such a farmhouse until one has made a large-scale plan of it on paper. Then things emerge that one hadn't suspected merely in walking round it.

Hatchetts, as she shows, is the product of no less than four building periods, or five if we include, as I think we should, the internal changes she and her husband have made since they took over in 1960.

First, there was a simple two-bayed 'hall-house', characteristic of the late medieval period of farmhouse building. Then it was extended early in the seventeenth century to provide a kitchen, and a proper fireplace and chimney. Previously the only fire had been on a central hearth in the old hall, which had therefore served as a kitchen as well. Building a separate kitchen was a big step forward in the evolution of the ordinary house. Many larger farmers were making such additions from about 1550 onwards. At the same time that the kitchen was added at Hatchetts, two bedrooms were also added—again a typical improvement of the time, with the growing number of children to be housed. Often a late Elizabethan or Jacobean family was as big as many a Victorian family. So Hatchetts had become a five-roomed house by say 1650 or a bit earlier.

The third period of enlargement and improvement was in the eighteenth century; and then small alterations were made about 1835, and the house was virtually complete, as we see it today. Four different periods of building, all reflecting some change in social history, and all employing local materials—white cob or puddled chalk for the first house, then timber-framing on flint foundations, then red brick. Finally, from 1960 onwards, come 'all mod. cons.'—hot water, a bathroom, mains water, and septic tank drainage.

The farm itself seems to have changed a good deal, having lost (I would think) most of its land. Now it is down to 12 acres. There is evidently a good deal to be discovered about the farm and the old fields, but for the time being we must be content with a nice account of the way in which a Hampshire farmhouse has evolved from a simple two-bay dwelling to the much more complicated house of today. Many English farmhouses, and I think even more in Scotland, were rebuilt all at once, in some period of prosperity, but others have 'just growed' like Hatchetts, bit by bit, over four different centuries.

HATCHETTS FARM, NETHER WALLOP, HAMPSHIRE
by Dorothy Beresford

The Wallop Brook wells up from a spring in Over or Upper Wallop, increasing in size as it runs along the valley between the downs to Lower or Nether Wallop, thence on to Broughton and Bossington where it joins the river Test, so famous for its trout fishing. Seen from the air the wooded valley which shelters the village of Nether Wallop runs like a dark ribbon of green through the lighter green fields which stretch out on either side of the valley. A mile to the north-east of the church is the ancient hill fort of Danebury, or Dunbury, the finest hill camp in Hampshire, which dominates the surrounding countryside with its crown of beech trees 150 ft. above the level of the plain around.

Camden tells us that 'Wallop' means 'a pretty well in the side of a hill'; Ekwall says that it derives from the Old English *wiellhop* meaning 'valley of the stream'. Dr. Hoskins says that Walla Brook (on Dartmoor) was originally *weala Broc*, 'stream of the Welsh or Britons', and the likeness to Wallop Brook is so striking that I feel that this could be its real meaning, a name given by the Saxon settlers to groups of older inhabitants living here when they arrived.

In Guest's *Chronicle* we read a translation from Nennius (in Latin) about a battle between the Saxon king Vortigern and the Romano-British king Ambrosius in A.D. 508 when the latter was slain by the Saxons; it is spoken of as *Guoloppum* or the battle of Guoloph. Nearby, the Roman road runs from Old Sarum to Silchester and below the hill fort on Quarley Hill lies the tract of country known as Wallop Fields, where legend has it a battle took place. It would seem possible that the Wallop Valley has been farmed from the time of the Belgae, *c.* 300 B.C.

Hatchetts Farm is just 1 mile from the Norman parish church, itself built on Saxon foundations. The farm had the stream as its boundary on its south-west side, and running parallel with the stream 220 yds. of footpath, which later became a carriage way, and is now a very busy road. The small 12-acre farm as it exists today, a neat oblong enclosed by hedges, must be roughly the farm of the oldest part of the house (early sixteenth century), except that the frontage to the stream has been lost through squatter's rights.

Many people in the village remember the kitchen garden which bordered the stream early this century and the apple trees, wild raspberries and friendly alien plants bear witness to this. The hedges surrounding the rectangular field contain hawthorn, black-thorn, spindle, wild rose, blackberry and elderberry, all firmly established. The soil is chalk, with light top soil and rough pasture in the field, which is well-drained, being above the roof-level of the house and part of the valley escarpment. A large dell in which grow 300-year-old yews with hollies and sycamores, which is now used as a

5. Liberty Farm, Halstock

6. GODLINGSTON MANOR, SWANAGE

scenic feature of the garden, must have been the chalk pit from which the first house was built. Next to the 'dell' is another smaller disused chalk pit, probably used for chalk which was slaked and then used for top dressing the soil. There is very little soil in the dell or on this north side of the stream but on the south side of the stream the soils is black and rich. The remains of a 'port sluice' opposite the house in the stream show where the stream was dammed in order to flood the water meadows on the south bank of the Wallop Brook so that they produced an extra crop of hay each year. This was still done in living memory.

The prevailing south-west wind blows along the valley, and the builders obtained maximum shelter from this by building the house end-on to it, and the large barn for cattle sideways to the wind, so that the central courtyard, or cowyard, now a garden, gets the most shelter.

The house is a long, low, thatched farmhouse which has gained its present size and shape as a result of four periods of building. It started life as a 'two-bay hall house', and the 'hall' of that house remains much as it was when built, except that it is now a coal house! It is built of fine white 'cob' or puddled chalk, dug from the chalk pit at the back of the house. The two-bay house had one door, and still has only one, on the court-yard side, and the worm-eaten oak lintel could well be the original. There is a small window in the south-west side overlooking the stream and the cart-track running parallel to it. The hall is $16\frac{1}{2}$ ft. (one rod, pole, or perch) from door to window, and open to the roof. Birds still fly in under the thatch. Until 1950 it had an earth floor, in which springs well up in February and March, but it is now covered with paving stones. When first built there would have been an open hearth centrally placed and the smoke filtered out through the thatched roof or under its open eaves. The walls of hall and bay are 9 ft. high. A door leads into the bay which has a very small window in the south-west wall, while above it was a very ancient oak plank floor (so full of death watch beetle that it had to be completely removed in 1960). In the original farmhouse this must have been reached by a ladder and was used as a loft for sleeping in. At the north-east corner of the house was a well, since filled in, but still in use in 1961, worked by a small electric pump over the brown glaze sink. I am told that a medieval farmer was allowed 6 acres of adjoining land for every bay of the house he owned, so a two-bay hall house would entitle him to 12 acres of land, which is the size of the farm today, if one includes the now lost stream frontage, into which one of the house drains still runs! I would guess that this part of the house could have been built as early as 1500.

The first addition to the house must have been made in the early seventeenth century. The eastern chalk wall was replaced by a timber-framed wall with wattle and daub infilling, the wattle used being the older type of hazel twig, not wooden slats, and the south-west, north-east and new east end wall were all built on flint foundations with Hampshire style brick and flint walls. This addition was built to provide a kitchen, and was also 1 rod in width. The chimney, still intact, is a large one, open to the sky, Tudor-style, and still contains the iron rods used for hanging the sides of bacon on for

E   65

curing, and has nails up the chimney which I imagine were used by the child chimney sweepers. Unfortunately the fireplace has undergone so much and varied alteration that it is very difficult to get a true picture of it; but from our deductions of fragments left, it was built with a Tudor fireplace, hearth and inglenook, which had its bread oven and its summer oven of which traces remain in plaster, and brick. The chimney stack measures 7 ft., and is $4\frac{1}{2}$ ft. thick.

In the parish register we read of a Nicolas Hatchett marrying a Maria Gore, in 1633. The Gores were well-to-do people in the fifteenth century, so from this it would seem that the Hatchett family were among the 'upper crust' of the then village. The chalk wall which divided the original 'hall' from its bay was extended to the roof by timber framing and brick infilling, thus making a bedroom above the bay; whilst above the new kitchen was another bedroom leading into the other old one. This new bedroom had a small casement window with leaded silk-glass panes, which is still there today.

So by the seventeenth century Hatchetts Farm was the home of yeoman farmers, with the old hall, a bay, a kitchen, and two rooms above. A door opened from the kitchen to the south-west and another on the north-east to the courtyard. The whole still thatched. The bedroom floors were of stout oak boards laid on the beams of the kitchen ceiling, and are still there.

By the eighteenth century there was again need for enlargement. This addition was built of red brick, continuing the line of the house, the rooms still measuring 1 pole from south-west to north-east, and an extension of 22 ft. was made. The entrance to the new room necessitated cutting through the tie-beam of the outer eastern wall, to make a door. From the position of the roof timbers it seems that the room above was only 14 ft. long, so that the eighteenth-century house had a roof which dropped to the level of the first floor ceiling. The ceiling of the new room was plastered, and the floor was earth, covered with boards, as it still was in 1960. The room has two casement windows on the south-west side, and one on the north-west wall; there were two doors, one leading from the kitchen, the other leading upstairs to the bedrooms, and each bedroom (now three) led into the other. There was no door leading outside from this room.

Hatchetts were living here throughout the eighteenth century. We read in the Inclosure Award of 1797 that James Hatchett had awards of 'allotments' which, without the 12-acre field in which the farmhouse is, totalled more than 150 acres. Land Tax assessments for 1800 show some Hatchetts lands assessed at £1 1s. 8d. and 16s. 11d. In 1827 Joseph Hatchett died aged 82, and he would appear to be the last male Hatchett who lived here. In the parish register we read that Stephen Webb and Mary Hatchett of this parish were married on 18th April 1811, and by the 1830s the Webbs were living at Hatchetts. Old people of the village still remember the Hatchett-Webbs living here. They say, too, that the Webbs were cousins of the Hatchetts, so the farm continued to be in the family but through the female line.

I think it is pretty certain that about 1835 the last alteration to the exterior of the house was made, when the east end roof was raised to the level of the rest of the house, and a chimney was built into this wall which served two fireplaces, one downstairs and one in the new bedroom or dressing-room. (The chimney has since been removed and the fireplace become a bookcase!) Neatly incised on two bricks in the east front of the large barn is the date 1835, which dates the renewal of the chalk wall of the large barn and its replacement with weatherboarding, which is still there. Also about this time, in the parish register, Stephen Webb is listed among 'substantial householders' of the parish who are subject to the parish rate of one penny in the pound.

The house now looks much as it looked in the 1830s, except that it has been white-washed outside to make it an entity. In the last decade two windows and a door have been added to the east wall, to give more light indoors. The whole is still thatched.

There is a large barn, 92 ft. by 19 ft. (outside measurements) which faces east and west, with fine roof timbers, and seven bays. Large doors in the middle open both sides and would have been opened during threshing to allow the prevailing south-west wind which blows along the valley with force to winnow the grain which was threshed on the floor of the barn. At one end is a large loft with a door for loading hay. The original barn was of cob (chalk) but the front has been replaced (as I have mentioned, about 1835) with weatherboarding, on a brick and flint wall base.

A small portion of the thatched or tiled wall only remains, which wall encircled the courtyard or cowyard, but chalk walls need considerable attention and the greater part of it had fallen down. The flint wall base remains and is planted as a rockery. The thatch has been replaced about 30 years ago by corrugated iron roofing on both barns.

There is a calving barn also of similar structure. On the roof supports are dates written in copper plate hand from 1833 to 1889. This barn has a floor above, in which the cowman or shepherd slept during calving or lambing.

There is a granary which experts date about A.D. 1700. It is built on staddle stones to keep the rats from entering, made of weatherboarding with a red-tiled roof. It has a pigeon loft with six entrances for pigeons, and a tiny window for light. Cart sheds and stable adjoined this within living memory, but now are demolished and a kitchen garden takes their place. Here too is the orchard which still contains some very old apple trees; and an ancient horse chestnut tree (estimated 300 years old) has given many cattle shade.

The Hatchetts had several fields and enclosures in the parish before the Inclosure Award, and kept sheep on the common downs and fields. They also had lands called Short Dagg, Out Fields, Gore-Land and home fields; this in addition to the present Hatchetts Farm.

The life on this farm must have continued to a very similar pattern from the time of its original building, from about A.D. 1500 to World War I. In 1912 Lord Bolton who had owned the lands in the village since 1840 (further back I cannot go) sold the estates in several lots. Hatchetts Farm was sold to a peasant farmer, Mr. Rawlings, with farmhouse, barns and its 12 acres of land, plus the stream frontage (then a kitchen and fruit garden but since lost to Hatchetts through squatter's rights) for £500. Mr. Rawlings moved out in 1952, and several townspeople moved into it for short spells. In 1960 my husband and I moved into it, and we have made several alterations inside the house, bathroom, hot water system, mains water and septic tank drainage, but we have tried to keep the character of the farmhouse. It is now a very desirable residence, the cowyard is a garden, the orchard still in use and the large field is kept in good condition by supporting a small flock of 50 ewes and their lambs, about 120 sheep at the moment. So the tradition of sheep farming in this part of Hampshire is continued. Tithes were redeemed completely in August 1959, and had been £1 4s. per annum.

In the present telephone directory the name 'Hatchett' occurs only twice; The Hatchet Inn, at Chute, some 10 miles distant; and E. E. Hatchett, farmer, in Winterslow, some 5 miles distant. No Hatchetts live in Nether Wallop now, and many of the fields which were farmed by Hatchetts in past centuries are now swallowed up by Middle Wallop Airfield.

A Gloucestershire Farm

Hunt Court Farm at Badgeworth, near Cheltenham, has a very long history, and the following essay on it has been severely reduced in order to make it manageable for this book. What has been selected for fuller treatment is designed to show the unusual records that may be available for a farm-history, besides all the more obvious ones. Thus the cash-book or accounts of a big farmer—Thomas Hinson of Hunt Court—running from 1611 to 1647—would make a marvellous book for publication, judging by the brief extracts given here. It is rare for us to possess such a complete picture of the day-to-day working of a farm at this period.

The other record which has been chosen for full treatment is the accounts of a farm-sale at Hunt Court in the spring of 1831, containing voluminous details of farm stock, gear, and implements, and the prices realized, and also all the household goods—a complete description in its way of a late Georgian farmhouse of some substance. There are tens of thousands of inventories for Tudor, Stuart, and early Georgian farmers, mostly now housed in county record offices, but these generally peter out round about 1730, so that nineteenth-century inventories like this are particularly valuable. For those who are interested in antiques the prices are mouth-watering today.

All such records should be carefully preserved even though they may not appear to be of great antiquity. Even the records of the early twentieth century are now of historic interest. The years before 1914 were another world, and some of us are beginning to think the same of the years before that other great landmark, September 1939. It is hard to throw away almost anything, but if room is needed then the local record office should always be consulted before anything is destroyed, however trivial it may seem to be.

*

HUNT COURT, GLOUCESTERSHIRE
by J. Ganley

Hunt Court lies 200 ft. above sea level in mid-Gloucestershire in the parish of Badgeworth, north of the main A 46 Stroud to Cheltenham road and overlooked by the Cotswold Hills. The present farmhouse stands at the foot of a small rise, the fields above running up to and forming the parish boundary with Brockworth. The name Court suggests that the farm was once a demesne farm, i.e. the home farm of the lord of the manor. 'Huntescourte' is first recorded as a name in 1422, probably named after an earlier owner or tenant.

A very rewarding and valuable find from the Gloucester Reference Library was a cash book of Thomas Hinson of Hunt Court, Badgeworth from 1611–47. The book contained rents received, wages paid for haymaking, sales of wood and notes of tithe corn and hay. Unfortunately it is rather difficult to read for the untrained eye, but some information can be obtained. He paid his servants as follows:

	£	s.	d.
Thomas Shayle	2	0	0
Will. Little	1	15	0
Henry Turner		8	0

His haymakers in 1630 were Goodwife Wilce who worked for 5 days and received 2/6, Goodwife Little 4 days for 2/– and Goodwife Duberley. Their husbands worked as well, one of them John Wilce earned 'for mowing Dengeworth and other grounds 7/–'. The days that they attended are ticked off and in all 8 men were employed, the married ones' wives helping out. In all it took them four weeks to complete the job. Draught oxen were used in 1633 for we learn that he paid his oxman 15/–. The book contained many pages relating to wood sold. In 1616 William Rodway paid 24/– for 22 lugs, and Thomas Cox paid 21/– for 22 lugs. In front of the book there is a note saying 'my best pyed mare took to horse today'. A family bereavement occurred on 10th August 1667 for he noted that his grandfather died.

Mr. Hinson was a generous employer for he gave his employees New Year gifts in 1630:

> T. Buckes wife a baskett of apples.
> W. Little a cuppell of capons.
> Spiche a baskett of apples
> Sallow a cuppell of rabbitts.

He cut 13 acres of wheat in 1630 and paid 3/– to his reapers for mowing barley and 2/– to 3 women for raking. The tithe paid was 20/– in 1637, for in his book he says 'My brother Gwinett Church Warden for my taxation of Badgeworth Church for this present year being xx^s taxed'.*

Both the fields Newleaze and Dengeworth are mentioned by name in 1633. The rents were collected from his tenants at Michaelmas and Lady Day and full lists were given for various years. The figures are in Roman numerals and certainly require much studying to understand fully. Sometime in the 1620s his tenants paid him the following rents and it seems the crosses written on one side could have been the tenants' mark to say that they had paid.

* This is more likely to be the church rate, rather than tithe (W. G. H.).

A GLOUCESTERSHIRE FARM

John Buchell	10/10
John Caper	14/–
John Bridgman	5/–
Widdow Merry	10/–
Rob. Edwards	3/4
Will. Longre	2/–
Penexton	2/–

Amongst his notes Mr. Hinson states:

'I payed Cantell and Ropson both when they went thyther 8d.

I gave hym 10d. to carry hym and his horse thyther.

I gave Elyon the long man to go with hym and to bring home my horse 10d.

I gave to Goodman Cofe for tyndnge (tending) my cattle for the week 4/–.

To the carpenter for work done 23/–.'

The Hinsons still owned Hunt Court in 1677 for in the Hearth Tax assessment of that date Thomas Hinson was taxed for 6 hearths which must refer to the original house on the moated site. By 1744 the Hyetts had become owners of Hunt Court.

A very interesting find in the County Record Office, in papers belonging to the Hyett family, is a notebook relating to a farm disposal sale held on 23rd and 24th March 1831. This sale was brought about because of the death of the tenant Mr. Herbert and it gives us a fascinating picture of the past farming techniques and prices of stock. The following gives details of the sale.

COWS

	£	s.	d.		£	s.	d.
12 Cows in/with calf	147	15	0	1 3-yr.-old bull	10	0	0
11 3-yr.-old cows	132	15	0	1 2-yr.-old bull	3	17	2
6 2-yr.-old cows	37	0	0	1 1-yr.-old bull	5	15	0
6 Yearling heifers	28	5	0				

HORSES

	£	s.	d.		£	s.	d.
1 6 yr. cart gelding (Boxer)	20	0	0	1 6 yr. cart gelding (Short) (one eye missing)	8	5	0
1 6 yr. cart gelding (Blackbird)	22	0	0	1 6 yr. cart gelding	10	15	0
				1 Cart mare	6	0	0

PIGS

	£	s.	d.
4 Store pigs	2	4	0

HAY

	£	s.	d.		£	s.	d.
10 tons in stack yard	29	8	0	10 tons in orchard	4	10	0
10 tons in rick yard	17	10	0				

IMPLEMENTS

	£	s.	d.			£	s.	d.
3 Wagons ¾ bedded	27	5	0	1 Crosscut & handsaw			11	6
1 Light wagon ¾ bedded with				Quantity sheaf pikes & rakes			17	0
iron arms	3	17	0	Quantity carpenter's tools			8	0
1 Straight bed & two wheels	1	5	0	1 Long ladder			3	0
2 Strong broad-wheeled carts	15	0	0	2 Ladders			7	0
2 Long narrow-wheeled carts	7	9	0	1 Iron bar	1	4	0
6 sets Long gearing	9	12	0	2 Sheep racks			2	0
2 sets Thillers gearing	2	0	0	1 Sheep rack and 1 calf rack			2	0
Assorted horse gearing	2	0	0	4 Cow cribs			3	6
4 Long ploughs	4	5	0	6 Cow cribs			4	0
4 prs. Harrows	5	0	0	Quantity Hurdles		2	15	0
1 Barley roller	2	0	0	Quantity flake hurdles			2	3
1 Large dray	1	18	0	12 Hair cloths		1	10	0
2 Plough drays		17	0	1 Grind stone & frame			4	0
1 Winnowing machine	4	0	0	1 Grind stone & frame			1	3
Assorted sieves, bags, shovels				1 Tallet ladder & 2 benches			10	6
Quantity bushell measures				1 Goose coop & hen coop			2	6
Quantity rakes, ropes, ladders				1 Goose coop & hen coop			1	0
1 Chaff machine by Passmore	3	0	0	1 Goose coop & hen coop			1	0
1 Chaff box & knife		9	6	1 Lead pump with pipes		2	10	0
1 Bruising Mill for malt and				1 Long oak pig trough & spout			15	0
corn, with wheel, strap				2 Pig troughs & long ladegaun			1	9
hopper	4	0	0	1 7-ft. Cider mill & press		TO BE LEFT		
1 Malt mill		3	0	7 Stone staddles & frame		1	12	0
1 Bean splitting mill		4	6	7 Stone Staddles & frame		1	8	0
2 Wagon ropes	1	3	6	7 Stone staddles & frame		2	4	0
Quantity draing tools		8	0	7 Stone staddles & frame		2	0	0
Beetle and wedges		8	6					

DAIRY

	£	s.	d.			£	s.	d.
1 Capital double cheese press	4	8	0	2 Milk skeels			11	0
1 Heavy cheese press		4	0	2 Milk skeels			15	0
1 Heavy cheese press		1	6	1 Whey skeel			11	0
1 Large milk lead	1	4	0	Pail, milk buckets & yolk			8	6
1 Whey lead	1	7	0	Pail, milk buckets & yolk			8	0
Barrel churns		13	0	1 Milk pail, 2 sieves, 4 butter				
Barrel churns	1	11	0	boards			8	6
1 Cheese cowl, tram & ladder		9	0	Cheese vats with suits &				
1 Cheese cowl, tram & ladder		7	6	hoops			10	0
1 Cheese cowl, tram & ladder		9	6	Cheese vats with suits &				
1 Large butter skeel		5	0	hoops			10	6
2 Milk skeels		17	0	Cheese vats with suits &				
2 Milk skeels		13	0	hoops			6	6
2 Milk skeels		10	0	4 Milk pans & 4 cream pans			3	6

DAIRY—*cont.*

	£	s.	d.		£	s.	d.
2 prs. Butters & milk tin		3	0	2 Large brass milk kettles		15	0
1 Dresser & 12 cheese skeels		10	0	1 Large iron beam with scales			
2 Benches & chopping block		10	6	& weights	1	11	6
1 Large brass milk kettle		7	0	1 Large meat safe		15	0
1 Large brass milk kettle		10	0	1 Salting tub		12	6
1 Large brass milk kettle		10	0	6 Frames with wire lattice		5	0
1 Large brass milk kettle		4	2				

CELLAR

	£	s.	d.		£	s.	d.
2 Wellbound hogsheads	2	18	0	1 Wellbound Hogshead & pipe		18	0
2 Wellbound hogsheads	2	4	0	1 Seasoned half hogshead	1	2	0
2 Wellbound hogsheads		10	6	2 Seasoned half hogsheads	1	6	0
2 Wellbound hogsheads	1	12	0	2 Seasoned half hogsheads	1	2	0
2 Wellbound hogsheads		14	0	2 Smaller Casks		6	6
2 Wellbound hogsheads		16	0	2 Smaller casks		15	0
2 Wellbound hogsheads	1	10	0	2 Trams			9
2 Wellbound hogsheads		16	0	2 Coffers		3	0

KITCHEN

	£	s.	d.		£	s.	d.
2 Brass candlesticks		2	6	1 Knife box & ass. knoves		8	9
2 Brass candlesticks		4	3	Window curtains & rod		1	9
4 Iron candlesticks		1	6	1 single-barrel gun		11	6
4 iron candlesticks, 2 lanthorns		2	0	1 Single-barrel gun (Ryan &			
2 prs. Snuffers treys		4	3	Watson)	1	6	0
1 Brass mortar & pestle		1	9	1 Single-barrel gun with add.			
1 Brass fender & steel tripod		1	0	barrel	1	0	0
1 Copper warming-pan		4	9	11 Pewter dishes, 32 plates &			
1 Cleaver, grid iron, dredger,				porringer	1	8	5
pepper box & toasting				Quantity pewter	NOT SOLD		
fork		3	9	1 Large frame table with oak			
1 Spit cauldron, dripping-pan				top & bench	1	0	0
& pudding-pan		2	3	1 Oval oak table		12	0
1 Spit cauldron		1	0	1 Round table		2	9
1 Fender, 2 prs. tongs and fire				8 Frame chairs	1	10	0
shovel		3	6	1 30-hr. clock in oak case	4	0	0
1 Copper tea kettle		9	6	1 Painted dresser with shelves	NOT SOLD		
1 Copper tea kettle		1	9	1 Painted cupboard fitted in			
1 pr. Flat irons		2	3	recess		12	0
2 prs. Flat irons		2	0	1 Smoke jack & chain	1	2	0
1 pr. Hand irons and bellows		3	6	2 Flitches bacon	2	2	6
1 Salt coffer		3	0	Quantity cups & jugs		1	6
2 Salt coffers		3	9				

LARGE PARLOUR

	£	s.	d.		£	s.	d.
6 Mahogany chairs satin hair	1	13	0	4 Pictures with gilt frames		10	0
11 Mahogany chairs satin hair seats		11	6	1 pr. Cut quart decanters		9	0
1 Square oak dining-table		19	0	Tableware, glass & 1 rummer		4	6
1 Square oak dining-table	1	0	0	1 set Casters in morocco frame		5	3
1 Dial barometer		16	0	3 Cut glass salts		9	0
1 Wire fender with brass top		6	6	2 Corkscrews, jappanned waiter & mahogany waiter		3	9
1 Double oak corner beaufit	1	14	0	1 Oval mahogany tea tray		3	9
Festooned chintz window curtains		9	0	2 Jappanned tea trays & waiter		1	0
				1 Jappanned tea urn		6	0

SMALL PARLOUR

	£	s.	d.		£	s.	d.
1 Square oak table on double pillar		5	6	4 Frame chairs		13	0
1 Painted writing desk		3	0	1 Picture & map			4
				1 Wire fender & brass top		2	6

CHAMBER OVER PARLOUR

	£	s.	d.		£	s.	d.
1 Four-post bedstead with chintz furniture	4	8	0	1 Mahogany night commode		14	0
2 Blankets & bed quilt		7	9	1 Blue & white chamber service		2	0
1 Seasoned feather bed & bolster				6 Chairs with seg seats		13	6
1 Straw palliasse		8	0	1 Coffer		1	9
1 Dressing-table & swing glass		2	0	1 pr. Demity window curtains & valance		3	6

CHAMBER OVER DAIRY

	£	s.	d.		£	s.	d.
1 Four-post bedstead with cotton furniture	1	4	0	6 Painted chairs with seg seats		15	0
3 Blankets		10	0	1 Oak linen chest		15	0
1 Seasoned feather bed & bolster	4	13	4	1 Oak linen chest		15	6
1 Tent bedstead with demity	2	12	0	1 Coffer		3	3
1 pr. Demity window curtains		3	0	1 Painted dressing-table		3	3
1 Oak chest drawers		14	0	1 Wash-hand service		4	0
				3 Carpets		NOT SOLD	

ROOM OVER BACK KITCHEN

	£	s.	d.		£	s.	d.
1 Oak stump bedstead & mat		8	0	2 Coffers		4	0
1 Flock bed & bolster	1	6	0	1 Cradle, 1 crib and child's chair		3	0
1 Bed quilt		3	6				
3 Blankets		11	3				

A GLOUCESTERSHIRE FARM

ATTIC CHAMBERS

	£	s.	d.		£	s.	d.
1 Stump bedstead		1	6	1 Flock bed & bolster		12	6
1 Stump bedstead		2	0	1 Bed quilt & rug		1	0
1 Stump bedstead		1	6	2 Rugs		3	3
1 Stump bedstead		1	6	2 Chairs		2	2

STORE ROOM (IN ATTIC)

	£	s.	d.		£	s.	d.
1 Large 3-fold clothes horse		5	6	1 Sway & bow fender		3	3
1 Large 3-fold clothes horse		4	3	1 Kitchen grate		6	0
1 Saddle & bridle	1	1	0	1 Bath stove grate		4	9
1 Saddle & bridle		4	0	1 Stafford grate & blunderbuss		1	3
1 Saddle & bridle		3	6	1 Small cash & pipe box		1	0
1 Side saddle	1	10	0	2 Pipe rings		1	6
1 Oak dresser, 3 drawer		10	0	1 Flight steps & ladder		10	0
Quantity Glass bottles	1	7	0	1 set Corner shelves		1	3
1 Polished kitchen grate		5	6	Quantity of old iron		11	0

LINEN

	s.	d.		
9 prs. Sheets	2	0	6 Towels [sold with pillow cases]	
5 Pillow cases	7	0	6 Table cloths [sold with pillow cases]	

BREWHOUSE & BACK KITCHEN

	£	s.	d.		£	s.	d.
1 Copper furnace, stack lid with iron work	1	5	0	2 Joint stools		1	6
1 Iron furnace	NOT SOLD			1 Ladegaun, large bar & tun pail		8	0
1 Iron furnace	NOT SOLD			1 Oval brass pot & lid		6	3
1 Iron furnace	1	3	0	1 Copper pot & lid		7	6
1 Copper furnace	1	7	6	1 Tin fish kettle 2 s/pans		1	9
1 Large mash tub & tram	1	1	0	1 Dutch oven, toaster & fr. pan		2	9
2 Wood skeels	NOT SOLD			1 Frying-pan, cheese toaster & s/pan		1	6
2 Tubs		5	6	Quantity tin ware		5	9
2 Tubs		7	6	1 Wire riddle, coal hammer & shovel		2	3
2 Tubs		4	0	1 Plate rack, dough tray & cover		13	0
2 Washing tubs		7	6	1 Mop box		3	6
1 Brewing sieve ladder		3	3	1 Dresser		15	0
8 Wooden bottles		9	0	Quantity earthenware		4	0
8 Wooden bottles		7	4				
8 Wooden bottles		7	6				

GARDEN

	£	s.	d.		£	s.	d.
1 Watering pan & hand glass		7	3	1 Breast plough ; 3 kypes	NOT SOLD		
1 pr. Garden shears & tools		13	0	Quantity potatoes	1	11	6
1 Garden chair on 3 castors		14	0				

	£	s.	d.		£	s.	d.
TOTAL for household goods	£85	15	9	TOTAL stock and implements	£629	0	0

Hunt Court, at the time of the sale, must have had a large market for butter and cheese, beer and cider, judging from the items in the catalogue. This seems to be its main source of income although a year before the sale, in 1830, there was 114 acres of arable, which included 40 acres of wheat, 20 acres of beans, 9 acres of peas, 17 acres of turnips and 6 acres of barley. The pasture amounted to about 130 acres, the remainder probably being fallow. In those days the turnips were presumably grown for stock feed during the winters, as we use mangolds and fodder beet today.

The Hyett family continued to own Hunt Court until 1911. An auctioneer's sale catalogue in that year shows that Hunt Court, together with Bouchers Farm, which was now a separate farm, were put up for auction. Hunt Court was then 361 acres. The dwelling-house was fully described and except for the kitchen which was fitted with a range, soft water tank and furnace, and the large cheese room, which is the present larder, all the rooms were much the same as at present. The buildings are also much the same, the only difference being the stabling which consisted of 3 stalls, 2 loose boxes, and a harness room. Both the buildings and the house are described as being substantially built of brick with tiled roofs.

The soil of the farm is medium to heavy loam, although there is a shallow but wide bed of sand running across Bold Furlong and gradually ceasing altogether in Dengworth. The boundary hedge, which separates the farm from the parish of Brockworth, contains seven species of shrub. This, following the 'vegetation theory' of one shrub per 100 years growth, could make this hedge 700 years old. It seems as though this has always been the natural boundary of the farm, the other boundaries being formed by the A 46 (the Fosse Way) and Normans brook.*

Until 1939 the stock relied on two brooks and numerous ponds shown on both the Tithe and more recently the Ordnance Survey maps. All the ponds have now been filled in and so has the moat. In this respect this may make for more profitable and modern farming but certainly less interesting history.

* As a result of many changes since 1918, the farm is now only 100 acres. In 1838, when the Tithe Award was made, it was 458 acres. The biggest single change was in 1939, when 229 acres were sold off; and in 1949 a further 144 acres. This left the house and 85 acres, to which 15 acres were later added.

Four Yorkshire Farms

A Yorkshire Wold Farm

The large Wold farm described by Mrs. White presents several interesting and unsolved problems. These are mainly (but not entirely) archaeological and if we knew the answer to these we might know a lot more about the history of Bartindale Farm. The east Yorkshire wolds are criss-crossed with ancient dykes and 'intrenchments' of unknown age. Mrs. White refers to the Argam Dikes which form the entire eastern boundary of the farm; but a good deal of the original northern boundary also is marked on the map as 'Intrenchment'. It runs as far as the present north-south road, and I wonder whether at some date it joined up with Argam Dikes at Bartindale Plantation. We do not know the age and purpose of these dikes, but they are most likely to be boundaries; and if I may hazard a guess they may be the boundaries of a large Iron Age farm. It is significant that Mrs. White refers to Iron Age tools and arrowheads found on Bartindale Farm.

Again, what are the squarish enclosures she refers to north of the farmhouse, which are also marked on the 2½-in. map? Only excavation could determine their age and purpose.

As for the deserted village site, we know far less about it than we do of many. There are a hundred 'lost villages' in the East Riding, and the Wolds in particular are studded with these sites. One of the most spectacular is Argam, about a mile south of Bartindale, now represented by only one large farm, as the 'lost village' of Bartindale is also.

Bartindale is not separately mentioned in Domesday Book, probably because it is included under the mother-village of Hunmanby, but it had its own little church as early as 1115. We know next to nothing of when and why the village was deserted but it was probably in order to make way for large sheep-pastures in the fifteenth century or the early sixteenth. There is clearly a lot more history to be unearthed at Bartindale.

*

BARTINDALE FARM, EAST YORKSHIRE
by Joan White

Bartindale Farm is situated approximately 4 miles inland from the North Sea, almost due west of Flamborough Head. The farmstead itself lies in the bottom of the north-south dale formed by the first fold in the Yorkshire Wolds from the sea, with the land rising to the hilltop on either side. It is between 150 ft. and 300 ft. above sea level.

Today there is very little shelter in this area of large arable fields. There are only two plantations of about 4 acres each. Both are on the boundaries. At one time, no doubt, the land was mainly covered in scrub and bush.

The Wolds are well known for their underground streams, often called gypses where they come to the surface (*gypa*—a spring). Near the pond there is a pump—long in disuse—which drew its water from one of these. Also, at the joining of four fields there is a dew pond.

The greater part of the farm is on chalk, with varying depth of soil, in places not more than 2 in. In the dale bottom, running north to south, there is sand and gravel. Up the hillside to the east there are found numbers of quite large cobbles which the people of the nearby Anglo-Saxon village used to build the walls of their houses. There is some flint on the hill to the west.

The 515-acre farm is in a ring fence, with the farmstead practically central. The public road, running more or less north-south cuts it into two not quite equal halves. The farmhouse, the hind's or foreman's house, and all the buildings are on the east side of the road. Two cottages, built around 1912–13 are on the west side. About the turn of the century, two fields (together about 45 acres) were added to the next farm. It in turn lost land to the next and so on, until just outside Hunmanby village two small farms were created.

Before that the boundaries to the north adjoined the land that had been the common land of Hunmanby. To the east it joins the boundaries of Reighton, and to the south and south-west the boundaries of the lost village of Argam; to the west the boundaries of Burton Fleming. About 150 acres on the far west side of the farm are in the parish of Burton Fleming (North Burton) and these at one time paid tythe, known as Queen Anne's Bounty, to the church of Burton Fleming.

The buildings now are typical of a large corn- and beef-producing farm, with large modern sheds, grain silo and bins, drier and so on. These have been added since 1955. Previously there were two open cattle yards, with a range of boxes between them, mostly built of chalk and cobbles.

The barn, cleared out by the local Young Farmers' Club for a barn dance, looks very old. At one time it contained threshing machinery, driven by a horse turning a wheel in the yard outside. Some of the roof timbers look like old ships' masts. (I think this occurs in many old farm buildings near the coast.) Stabling for about 20 horses forms the north side of this yard. At a guess this was built in the early 1800s, although I have only the appearance to go by.

Beyond the buildings is the foreman's house built about 100 years ago. This comprises sitting-room, kitchen, back kitchen, pantry and cellar, five bedrooms and bathroom. The two back bedrooms and bathroom were originally one long dormitory, containing several beds for the single men who lived in with the hind.

7. BARTINDALE FARM, HUNMANBY

8. Reynard Ing, near Ilkley

The hedges marked on the map (see key) appear to be of greater age than the rest. They are on raised banks and do not follow a straight line, nor are they cut by any other hedges at any point. Working on the vegetation theory, they contain in some places six different species of bush and up to eight in others. Thus they could be 600 to 800 years old.

The lane forming the boundary with Reighton parish is on part of the Argam Dikes. These are so old that no one as yet seems able to offer much explanation of them, but they seem to have been either some kind of defence system or else a cattle boundary. They have been traced for many miles.

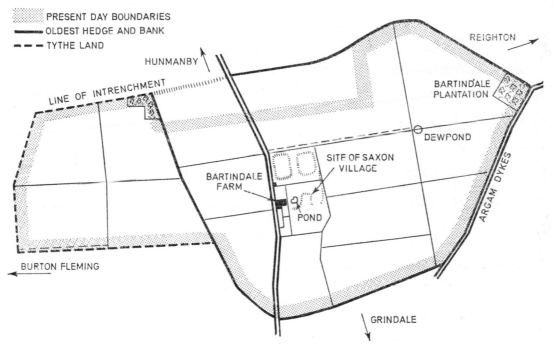

FIG. 6. Bartindale Farm, Hunmanby (East Riding)

The straight hedges, all of quickthorn, do not cut through this boundary at any place. They were probably planted after 1773 when a Bill was passed in Parliament which provided for the 'adoption of any scheme of husbandry that might tend to increase the productive power of the land, subject to a three-quarters majority of the occupiers of common arable land being in agreement with and having the consent of the owner and the tithe owner.'

Not much can be learnt from the field names. Each field is known by either its position, such as Stone Pit Field, or by its acreage. The name of the farm itself, variously spelt Barkedale (1270), Berkildale (1285), Bartondaill (1518), means 'Barkil's valley'. Barkil or Berkil is a Scandinavian personal name.

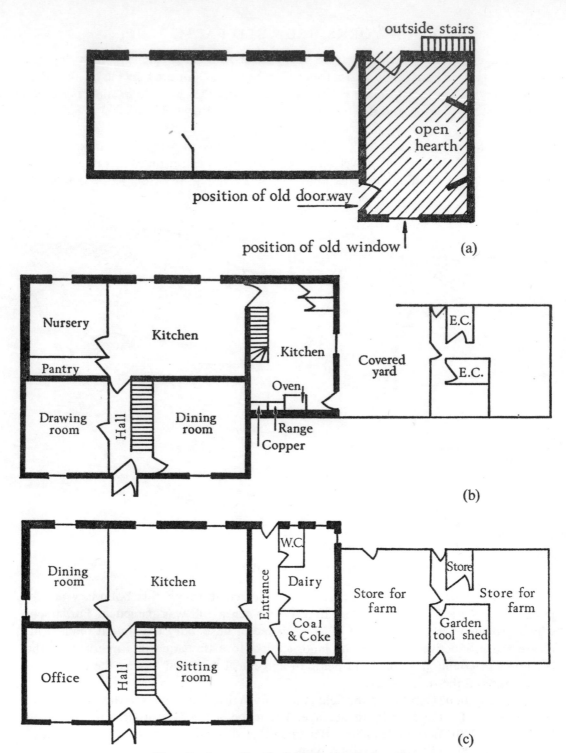

outside stairs

open hearth

position of old doorway

position of old window

(a)

Nursery

Kitchen

Pantry

Drawing room

Hall

Dining room

Kitchen

Oven

Range

Copper

Covered yard

E.C.

E.C.

(b)

Dining room

Kitchen

W.C

Entrance

Dairy

Coal & Coke

Store for farm

Store

Store for farm

Garden tool shed

Office

Hall

Sitting room

(c)

FIG. 7a, b, c. Bartindale Farm: house-plan

The foundation of the oldest part of the house was probably built at the same time as the Anglo-Saxon village. In 1962 when all this part was stripped internally, the floor had to be raised 1 ft. to make it level with the rest of the house. We found, on lifting the old floor bricks, that the floor had already been considerably raised; and we found an old lintel and blocked-in doorway behind the wall plaster. The top of this doorway was less than 3 ft. from the level of the brick floor, so the previous level must have been at least 3 ft. below that. This bears out my theory that the foundations are of an old village house, as the floors of Saxon houses were below ground level, both for warmth and shelter and also to save on building high walls.

The back staircase was on the west wall of this portion, but had obviously been put in at a later date than the beams of the floor above, as they had been cut through to allow for it. The stairs had originally been outside, as again, when the plaster was removed the position of the entrance was clearly visible. When the old sink was removed on the east wall, an open hearth was revealed. It was 10 ft. across with five flues opening into it. A window had, at a later date, been let into the back of the chimney.

The oven on the south wall (large enough to hold a man) was also a later addition. Behind it, when it was removed, was a blocked-in window.

The plans show the additions made to the original structure, which is shaded on Fig. 7a. It was extended first to the west, and Fig. 7a shows the shape of the house as it was until rooms were added on the south side, probably in the early eighteenth century. The south and south-west walls are brick, and the chalkstone walls of the earlier structure have been bricked over to make the old and new look alike. Fig. 7b shows the house as it was described in an inventory of 1913; Fig. 7c as it is today. These plans demonstrate the different living requirements of the different generations.

The Lost Village of Bartindale and the Enclosure

The lost village of Bartindale lies to the east of the present house and buildings. Many of the typical long-house sites can be traced still in the mounds of stones and earth in the grass.

There is no mention in Domesday Book of the village, so at that time it can have been of little importance. It became big enough, however, to have its own priest and chapel before it was depopulated sometime in the sixteenth century. The church of Hunmanby, together with its chapels (of which Bartindale was one), was conveyed to Bardney Abbey in 1115.

It is a known fact that the Benedictine monks of Bardney Abbey had several sheep granges in this area. I think this village was more of a large farm, with sheep as the main industry, especially as the land on the hilltop to the west was known as the Sheep Walks. It is nice to think that, in a sense, the tradition still goes on. My husband is a well-known sheep breeder in this area, and has won prizes at the Royal and other shows with both the sheep and the fleeces.

The enclosure is to the north of the present farmstead, and it is very clearly marked. It is a complete square 130 yds. × 130 yds., surrounded by double walls which form a boundary 8 yds. across. To the east there is a definite opening or gateway. This may have been a cattle enclosure for the village or it may be of a much earlier date. As yet I have been unable to obtain any further information.

That people lived here very much earlier than the Anglo-Saxons is certain. A neighbour tells me that his grandfather collected Stone Age and Iron Age tools and arrowheads, some of which came from this farm. These are now, I understand, on record at the British Museum.

A Farm in Craven

I can add little to the admirable account given by Mrs. Mason of Reynard Ing in Wharfedale except to make some comments on the North Country word *mistal*, the cow-house which formed the north end of the homestead. This is a pure North Country word, and it is not even certain what its derivation is. In the West and South of England the word is invariably *shippon* or *shippen*.

According to the *English Dialect Dictionary* the word mistal for a cow-house derives from two Scandinavian words, mjø (milk) and støl (a shed, a place) so it means the 'milking-place'. But another great authority (the *Oxford English Dictionary*) disputes this and derives the word from *mix*—an old word for dung—and stall, so the 'dung-stall'. The use of the word *mix* for 'dung' is now obsolete, but is well recorded back to King Alfred's time. But the curious thing is that the earliest recorded use of the word *mistal* is as recent as 1673.

The word shippon or shippen (either spelling can be used, but the former is usually given) has a very ancient history. It occurs as early as about the year 900 as *scypene*. It has nothing to do with sheep, as is sometimes rashly supposed, but comes from the Old English word *scypen* meaning 'cow-house, stable' and is similar in origin to the Old English *scoppa*, meaning 'a shed, a booth' from which our modern word 'shop' is derived. So the humble shippon and the smart shop in town have the same root in the Old English language.

According to James Walton's book *Homesteads of the Yorkshire Dales*, published in 1947, the cow-stalls were known as shippons in Wharfedale and Craven and as mistals or booises in the hills to the south. In the northern dales the word used was 'byre'. The mistal or byre or shippon was always placed at the lower end of the long-house for obvious reasons. Such long-houses, which housed the family and the animals under one continuous roof, were once a common type in the North Country and the West Country, but are now becoming very rare. A number still survive on Dartmoor, in the far south-west of England, and the cattle are still brought into the shippon in bad weather on small farms, though not normally Do any of these old long-houses survive in use in the North Country?

REYNARD ING, NEAR ILKLEY
by Kate Mason

Reynard Ing, our farm, is situated at the extreme eastern edge of Craven in the Yorkshire Dales. This was, at least until the seventeenth century, remote and thinly populated country, whose inhabitants were, in the opinion of an eighteenth-century writer, 'barred up by trackless wastes and impracticable ways'. Long ago, perhaps 3,000 years, some neolithic hunter lost his beautifully fashioned laurel-leaf flint arrow-head, to be picked up by one of our family after a night of heavy rain; from the Roman fortlet of Olicana 2 miles away, the road to Ribchester passed along the edge of our fields, although no trace remains. The Saxons followed the rivers and founded and named most of our villages, while the Norse sheep farmers occupied the high ground, leaving us hundreds of minor place names and many topographical and agricultural terms. The Normans brought little new blood, only a change of overlord.

We are fortunate in having a long sequence of deeds going as far back as 1630, but the land was already enclosed at that date and what happened earlier we can only conjecture. The earliest deed names our farm as 'the Close called Reynard Ing alias Reynolds Ing'. Professor Smith in *The Place-Names of the West Riding of Yorkshire*, suggests that 'Reynard' comes from a Middle English and Old German personal name, Reginhart, but he did not know of our deed, and it is most probable that it was indeed the 'ing' or meadow made by Reynard as the deed suggests. It is not possible to fix a date for its making, only to suggest that for such a corruption to have taken place a considerable time must have elapsed, perhaps 100 or 200 years, suggesting that it was first cultivated in the early 1500s.

The site of the farm lies about 100 ft. above the River Wharfe on a gentle, north-facing slope. The soil, a heavy clay loam with underlying boulder clay, tends to be wet. It must have been covered with dense woodland—small oaks and other forest trees grow by the score in the grass every year and are destroyed by grazing. The farm is a mile and a half from the village of Addingham.

Many becks and trickles of water flow due north from Addingham Moor to empty into the River Wharfe, and water and woodland between them were hard obstacles to clearance. It is no wonder then that anyone who tackled the clearing should be remembered by name. He must have had a good eye for the possibilities of the land as it is obvious that the making of the four fields, now known as Reynard Ing, the Threaps, Lime Kiln Close and Cocken Flatt, was conceived and carried out as a whole in one operation.

The 'ing' or meadow which Reynard made had a small beck meandering through the middle which must have spread out to make a large marsh. This beck has been diverted at the top of the field by means of a bank, and a new course has been made that runs due north to the bottom of the field. Furrow marks and signs of an ancient headland where the stream once ran show that about an acre of land was ploughed for a considerable time. The field to the east of the meadow, now known as the Little Threaps, seems to have been mentioned in the first deed. It lies on a gentle slope with hedges on the east, west and north, and a wall on its southern boundary. The hedges have in them an average of six different species of trees and shrubs, giving a suggested planting date of 500 to 700 years ago.

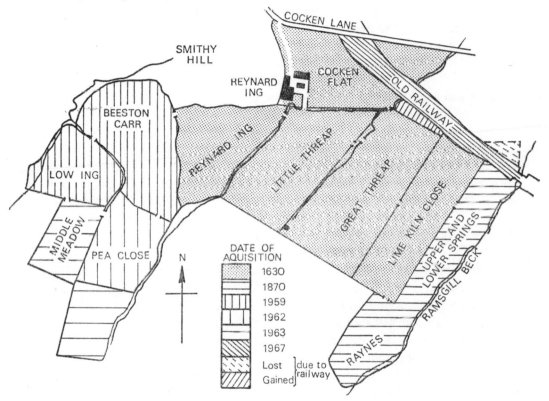

FIG. 8. Reynard Ing, near Ilkley (West Riding)

This agrees well with the suggestion of a possible fifteenth-century date for the clearing of the fields. This field of even slope, good shape and easy access, seems to have been a favourite one for ploughing but shows no ancient ridges or other marks.

The adjoining field to the east, now known as the Great Threaps and Lime Kiln Close, was divided into two during the eighteenth century. The Lime Kiln Close was first mentioned by name in 1795. There was probably a hedge to divide them, but this has now disappeared, though the remains of a few decrepit thorn bushes and alder trees were removed a few years ago. The field shows signs of a long period of cultivation; headlands and banks show that it has been quartered and each quarter has been ploughed separately. The land is now laid in narrow ridge and furrow ($7\frac{1}{2}$–9 ft. between one furrow and the next). These are very common locally and are supposed to have been an aid to drainage. They are known as 'six about lands'. The name 'Threaps' means 'disputed ground', but there is no surviving record so far as we know of what the dispute may have been.

The field in front of the house was also mentioned in the original deed. It is bounded on the north by a lane, now hedged. Although not mentioned in the deed, the lane is probably very old, indeed we know that the Roman road to Ribchester ran on approximately the same line. The field is known as the Cocken Flatt. 'Cocken' means cock or woodcock and is one of a number of pieces of land so called (Cocken Wood, Cocken Deep, Cocken End, Long Cocken).

The tenement which was built about or before 1700 remains in external appearance nearly as it was built. Typical of the time and district, it consists of a three-bay building facing east, with the house at the southern end, then stable and barn with winnowing door, the cowshed (or mistal) along the northern end. The stable later became disused as the smell penetrated the house, and in 1912 it, together with the stable loft, was converted into 'room' and bedroom for the house. The whole is built of random stone walling, using stones from the river which still show the moss upon them if a hole has to be made in the wall. The east wall is plastered with lime and cowhair and whitewashed.

The house itself, although with good-sized rooms, was small, only a 'house-place' and scullery or dairy, with corner taken off for a pantry, and stone-flagged floors. We surmise that there was originally only a loft above, reached by a ladder, the floor supported by three large beams and oak purlins which are open to the room below. The east-facing windows were (one still is) Yorkshire sash windows—that is the earliest type of sash window with one half sliding behind the other. The stone mantelpiece surrounds an open fire with iron reckon—or pot crane—for hanging pans. The fire would be of peat as there was the right of turbary on Addingham moor. At the side of the fireplace and built into the thickness of the wall is a 'beehive oven' which was heated by burning brushwood, raking out the hot ashes, and putting the risen dough to bake. Between the fireplace and the window, where is now a cupboard, was a stone slab, which latterly supported a washing-up bowl. This in all probability was a 'backstone' for baking haverbread—the North Country oatcake—which was the staple food of the district.

There is no room to tell of all the changes and developments brought by the eighteenth and nineteenth centuries, but it is interesting to take up the story again in 1899 when Charles Dilly, proprietor of an Ilkley hotel, took over the tenancy of the (by now) 30-acre farm at a rent of £90 a year. This was quite a high figure for the time which was one of depression, but he was an enterprising man and a good psychologist; he had taken over a bankrupt concern which he made into one of the best hotels of the time (1895–1915). He was of Irish descent and had a farming background, he loved a good light horse, and was a keen pig and poultry breeder. The farm improved out of all knowing during his tenancy. His chief concern was to have quality milk, cream, table poultry and eggs, and as much meat (bought on the hoof and finished on the farm) as was possible. All the hotel waste, and there was plenty, was fed to his 40-sow herd of Middle White pigs.

The Masons came into the picture in 1901. Mr. Dilly had sheltered with a shooting party, on the farm where Harold Mason lived. He noticed the careful and efficient way the lad fed his stock, and remarked that he would remember if he should need anyone to look after his stock. Harold and Agnes Mason were married on 1st June 1901, and lived in a small terrace house, while he was employed by a coal merchant, a job which they hated. A month later, Mr. Dilly's bailiff was killed in the farmyard. The promise casually made was kept, and the Masons settled in the farmhouse. They had free house, milk, eggs, and so on, and 25s. a week, and were considered very well paid. After seeing their furniture off from Ilkley, they walked the 2 miles with twopence in their pocket; at the gate they picked up a horseshoe, which still hangs behind the door.

In 1912, Mr. Dilly handed over the tenancy to his bailiff. During their office, 6s. a week had been paid to Agnes to board a farm lad, and she had the total sum in hand. The landlord, for a raise of £5 in the rent, provided just over £100 to pay for the conversion of disused stable and loft into parlour and bedroom, and to piping the water from the beck. All went well until 1920 when the farm was sold and the new owner gave notice to quit, after the stock had been sold, but before the notice expired the owner found himself in difficulties and the farm was offered at the same price, £2200, to Harold Mason, and he accepted.

This seems to be the place to cease the detailed story, for though history is continuous, it is too near and too familiar. It is sufficient to say that following the collapse of agricultural prices in the twenties, and the accidental death of Harold Mason in 1928, there were many difficulties, but from 1930 onwards things improved. We have erected, mostly by our own labour, some good and convenient buildings, and have added to the house without, we trust, spoiling its late eighteenth-century character. We have added 18 adjoining acres which enabled us to utilize a much better water supply and finally re-purchased the 2 acres of railway cutting, which the farm lost when the Skipton–Ilkley line was made in the 1860s. The cutting is being filled with earth and much indestructible industrial waste, and we have perhaps provided a happy hunting ground for some future archaeologist.

Two West Riding Farms

At first sight Wilstrop Hall and Hunshelf Hall seem to have little in common beyond the fact that they both lie in the West Riding of Yorkshire; and even here one of them (Hunshelf) lies high up the Don valley where steep-sided wooded valleys penetrate the Pennine foothills, and the other (Wilstrop) is right down in the Plain of York, not far from where the Nidd joins the Ouse and scarcely 50 ft. above sea level.

But they have certain other features in common, in their distant history at least, and this is really the only excuse for bringing them together like this. Both carry the name 'Hall', signifying the demesne or lord's home-farm at one time. Both were first settled well before the Norman Conquest and remained in English hands after the Conquest; and both (so I think) were once villages, the sites of which lie not far away from the present-day farmstead.

Hunshelf means '*Hun*'s shelf or ledge', a description that suits the site very well, as Mrs. Crossland says. In all probability *Hun* was the first settler on this rather difficult site. How far back he lived, we have no means of knowing. In 1066 the Saxon owner was one Alric, and though the estate passed to a powerful Norman—Ilbert de Lacy—Alric was allowed to stay on, possibly as the nominal owner under Ilbert. In many parts of England, the Anglo-Saxon owners were completely dispossessed by the Norman conquerors, but in parts of Yorkshire many were allowed to remain. Perhaps it was an attempt to pacify the North, as Yorkshire was involved in a massive rebellion against William the Conqueror in 1069. He repressed this rebellion, and others, with savagery, laying waste great tracts of countryside, which is why in Domesday Book (1068) Hunshelf is described simply as 'waste'. It had been worth ten shillings a year in 1066, but 20 years later it was worth nothing. But farming must have been resumed very shortly afterwards, and since then Hunshelf has had a continuous history as a farm.

One other point which Mrs. Crossland makes is worth taking up. She is almost certainly right in thinking that there was once a village near the present house. She mentions a field called 'Town Field', which she thinks may have been the land for the communal use of the people at that time. In fact I think this field almost certainly represents the side of the 'lost village' of Hunshelf. Down in the Midlands, where these lost or deserted villages are very numerous, I often found that their site could be identified by a field called 'Township Field' or sometimes 'Old Town Close'. I think 'Town Field' at Hunshelf is just such a clue, and would repay a close look for any signs of bumpiness or an irregular surface, especially one which cannot be ploughed because of stone underneath.

90

TWO WEST RIDING FARMS

Like Hunshelf, Wilstrop takes its name from the original Old English settler. It means 'Wifel's *thorp*', and 'thorp' has the special meaning of a daughter-settlement from an older village. It is possible that the mother-village was Kirk Hammerton, with which Wilstrop later appears to be associated; but I am more inclined to think that Marston (now called Long Marston) was the mother-village, as the river Nidd must have been a formidable barrier in olden times. And again, Wilstrop is grouped with Marston and Tockwith in Domesday Book, which suggests an original connection.

At any rate, Wifel founded a new settlement close to the banks of the Nidd, at some unknown date. In 1066 a large estate, which included Marston, Tockwith, and Wilstrop was held by a Saxon Aelfwin, who again seems to have been allowed to remain on, but under a Norman overlord—Osbern de Arches. Wilstrop Hall was the demesne farm of Aelfwin, as Hunshelf was of Alric.

We do not know the size of Wilstrop village in 1086 (any more than we do of Hunshelf) owing to the way in which Domesday Book was compiled over most of Yorkshire. We get our first idea of the size of the village in the early fourteenth century, before the Black Death, when, as Mrs. Blacker shows, it had some 28 farmhouses and cottages, a fairly large village by the standards of the time in the West Riding. The old village survived down to the closing years of the fifteenth century, when it was depopulated by the local squire—Wilstrop, who took his name from the estate. It is sometimes thought that the New Rich of the Tudor period were the principal offenders in bringing about the depopulation of so many English villages, but more often than not it was the old gentry who were rationalizing (as we should call it) their estates, and going over to cattle and sheep pastures rather than arable husbandry. It should be said, however, that in many cases that we know of the village population had fallen so low because of repeated epidemics of bubonic plague that arable husbandry was no longer possible. Pasture-farming came about partly because of higher prices for sheep and cattle and their products, but also because of the great economy of labour it entailed.

The old village of Wilstrop disappeared then about 1490, and the score or more of medieval farms are now represented by only seven farms scattered over the estate. The old village site is completely deserted.

*

HUNSHELF HALL (WEST RIDING)
by Phyllis Crossland

Our farm of 147 acres is situated in the southern corner of the West Riding of Yorkshire, 10 miles north-west of Sheffield, and close to the Pennine Hills. The lands contained in the fork of the greater and lesser Don form the township of Hunshelf, and there is much evidence to show that Hunshelf Hall which we now occupy as working farmers was owned in former times by people of high social standing in the district.

The name seems to be derived from some 'shelving' towards the Don. In Domesday Book it is written *Hunescelf*, meaning 'Hūn's Scelf', a shelf or ledge of land. Certainly the positioning of our land bears this out. Hūn was the Old English landowner here. The same name occurs in Hunslet, now a suburb of Leeds.

The site is a fairly sheltered one and although we are on a hillside we are away from the full force of the wind. A few trees strategically placed by persons long since dead act as a windbreak.

The soil is a light loam, and the type of farming mixed. We have dairy and beef cattle, sheep, pigs, and grow oats, potatoes and turnips.

We have had a mains water supply only since 1949. Before that date our water was pumped from a well situated at the far side of the farmyard near to the road but on the farm side of the wall. As there are some old steps leading over the wall from the road to the pump-house we conclude that this well, apart from supplying the Hall, must have been used also by the villagers of long ago. There is another well here, too, out in the front garden, covered now of course, but it was used by our predecessors for a time at the beginning of this century. They stopped using this well as the water hadn't as good a taste as that from the other well near the road.

The farmstead is fairly central in relation to its fields, but whereas now it lies half a mile west of the present-day village of Greenmoor, I am convinced that a much earlier Hall on this same site was once near the centre of a medieval village. I say this because, in addition to the fact that Hunshelf is mentioned in very early records, the tangible evidence here around the actual place is fourfold. Firstly, there are the steps leading to the well which have already been mentioned. Secondly, a large field of ours just across the road from the old well bears the name of Town Field. This could well have been land for the communal use of the people at that time. Thirdly, the stocks were situated at the roadside near the Town Field until 1937 when they were moved into the centre of Greenmoor and re-erected to commemorate George VI's coronation. Fourthly, there is a little old smallholding called 'Peck Pond' just along the road to the west, which was originally an inn known as 'The Brown Cow'. This building and one other very old house—Don Hill—are the only ones remaining now at our end of the present-day village. The people who used the stocks, the well, and frequented the Brown Cow must have lived four or five centuries ago or possibly longer, as their dwellings are no more. The reason for the shift in population was the stone-quarrying industry which developed during the nineteenth century near Greenmoor. Consequently the newer houses were built there to be nearer to the men's work.

It is on record that Ailric the Saxon held three carucates of land at Hunescelf before the Conquest, but they were returned as 'waste' in the Domesday Survey. It seems likely that a mesne lord was placed here in Norman times as we have information of two charters belonging to the reign of Henry III or Edward I. These state: 'Thomas, son of William de Hunescelf, gives to Elias de Walderscelf, a piece of land which Matthew de Hunescelf formerly held within the limits of Hunescelf. The tenant is to grind his corn at the mill of Hunescelf.' The second charter reads: 'Richard de Hundeschelf gives to Richard, son of Adam, and his heirs, pasture for all his cattle within the limits of the pasture of Hundeschelf, at the rent of *unum obulum argenti*.'* How long the Hunschelfs continued to hold the estate is not definitely known. There is no account of the course of descent which the manor took until it appears in the inquisition of Francis Wortley ancestor of the present Earl of Wharncliffe, in 1586. It could well be around this date that the older part of our existing house was built. It looks to be either late Elizabethan or early Jacobean. At any rate it is quite obvious to anyone, even with an inexperienced eye, that the present Hunshelf Hall comprises building of two distinct periods.

The old part appears to have been one large hall without upper floor to begin with. This floor, added later, 'cuts' the window in half, so that in the upper room the window is down at floor level. It has iron bars to it too. This old part of the house is gabled and the stonework more weatherworn than the newer part. According to the records, there was a fire at Hunshelf Hall in the early eighteenth century when most of the house was destroyed, together with a lot of the parish records. The old part left now is what survived the fire. We call it the back kitchen as my husband's family did use it quite a lot before I came here. It used to contain a big, old stone sink, wash copper and brew copper. We only use it as a store-room now, as really we have plenty of other living space. In the upper room the ceiling collapsed years ago leaving exposed to view all the ancient wooden timbers.

The greater part of the existing house was built on to the surviving part of the former one in 1746 by the owner at that time, a George Walker, and is typical Georgian in style. The builder's initials and date are to be seen clearly carved on a stone archway at the side of the house. This George Walker must have been an influential man of his time in the area, a sort of country squire. He is mentioned as encouraging the cloth trade to develop at Penistone, 4 miles away, by having a local Cloth Hall built instead of carrying cloth all the way to Sheffield. The clothiers had to make an agreement with him that: 'any person who took any kersey, plain, or other cloth to sell at Sheffield after 29th September 1743 should incur a penalty of three pounds for every piece sold.' Just when the Walker family took possession of the old house here, I don't know. This evidence could have been destroyed in the fire. We do know they had it as early as 1639, as seen in this interesting little bit of information from the Quarter Sessions accounts: 'On 16th January 1639 Elizabeth Pashley of Oxspring stole at Hunshelf a brass pan value six shillings and a small brass pot, the property of John Walker. She pleaded guilty and was fined 6d.' The last of the Walkers to own Hunshelf Hall were two sisters, spinsters, and it was then bought by a Mr. Smith at the beginning of the nineteenth century. When my husband came to live here in 1946 one of the bedroom windows was completely blocked up. This had probably been done in the early part of the nineteenth century to avoid paying the window tax. When my husband uncovered the window it

* A token rent of one halfpenny (W. G. H.).

was noticed that the glass was thicker than that in the other windows. It is the original 1746 window and has scratched on it *Dear Mis Walker*. The Mr. Smith who bought it after the Walkers let it to the Coldwell family who farmed here till 1904. They were in the 'gentleman farmer' class as can be gleaned from an original account written in 1912 of the early life of one of the old men of the parish, a George Marsh. He had told his story to the writer, making his mark X at the end, being unable to write. Telling of his boyhood in the 1840s and of his family's poverty, he said how they had been kept from starvation by the kindness and generosity of Mr. John Coldwell of Hunshelf Hall who often gave him food to take home. At the beginning of this century a service used to be held in the big room here on Sundays, taken by the Minister who used to come from Sheffield by train to Wortley Station to preach in the Chapel at Greenmoor. He would come on the Saturday evening and be 'put up' here for the night, then the family, servants, and farm workers would all assemble after breakfast for the service after which the parson would leave for the village and the Mr. Coldwell of that time would say, 'Now lads, back to work for you all'.

The Taylor family followed the Coldwells in 1904 and remained until my husband's family took it in 1946. In Mr. Taylor's day they still employed four men full-time besides seasonal helpers, and kept four working horses. Now my husband has only one full-time man and occasional casual help, but then he has two tractors and quite a lot of machinery. I think I am the first woman to look after this big house without help, but then I am the first to have electricity and mod. cons., so that explains it, and I don't make any butter or bake my own bread as my predecessors did.

The room we now use as a sitting-room was, to all appearances, a big living kitchen originally, with large open fireplace. This is evident as you look at the stone archways set into the walls round and at either side of the fireplace. This fireplace is a tiled, fairly modern-looking one, and had been put in before we came here, but it is not in keeping with the character of the room. If I could afford the money I would simply love to have it all knocked out and expose at least some of the original cavity, putting in a type of fireplace suitable for it. This room has two huge oak beams across it which fortunately haven't been plastered over. We have three other rooms on the ground floor, apart from the old back kitchen, two passages, two staircases, and five bedrooms and bathroom upstairs. There is a large cellar and a smaller one under two of the downstairs rooms, both with huge stone salting slabs, now obsolete, I'm afraid.

During our twenty years' stay here we have changed from well water to mains water, paraffin lamps to electricity, earth closets to flush toilets, tin bath to modern bathroom, and have become connected to the outside world by telephone and motor car. Yet, in spite of these necessary changes, this house retains much of its original character.

I would love to know more about the place, particularly when the older house was built, and I hope at a not too far off future date to be able to fill in some of the gaps in its history.

WILSTROP HALL (WEST RIDING)
by Trudy Blacker

Wilstrop Hall stands almost in the centre of its 365 acres of land. Perched high on the south bank of the River Nidd, it lies in the Plain of York, about 8 miles from York and 12 miles from Harrogate.

For centuries the Hall has been the head of the Wilstrop estate which, when it was last sold, consisted of 7 farms and 1,610 acres of land. The house stands on the site of an ancient mansion, once occupied by the lords of the manor, and a field near the house is the site of the deserted village. Although we have no village today Wilstrop is still a civil parish, attached ecclesiastically to Kirk Hammerton, some 3 miles away.

To the west and north, the winding River Nidd forms the farm boundary. The York and Harrogate railway line borders the eastern side and Wilstrop Wood forms the boundary on the south side. This is a very brief picture of our quiet farm today; now let us go back—a little over 900 years—and try to follow the fortunes of Wilstrop through the centuries.

The spelling of the name has had many variations. I found Willestrope, Willistrop, Woolstrop and Wilesthorpe, but—to make reading easier—I intend to use the modern spelling in this history. I found it impossible to separate the Hall from the rest of the estate until about the beginning of the eighteenth century so, until then, 'Wilstrop' must refer to the Lordship of Wilstrop, and not the actual farm, although this was the Manor House and the focal point of the estate.

I found my first reference to Wilstrop in 1066. After William the Conqueror's victory at Hastings he gave grants of land to many of the knights who fought so valiantly in his cause. One such knight was Osbern, and the land given to him included Wilstrop. Osbern took as his principal seat Thorp Arch, not far away, and took the name of Osbern-de-Arches.

In 1306, Robert de Pontefract died possessing lands in Wilstrop. The Exchequer 'extent' recorded that he had 130 acres in arable fields, 10 acres of meadow, some pastures in closes, an oak wood, a water mill, and a fishery. From the peasant-tenants he claimed a total rent of 4s. 7d., and the obligation to do thirty 'works' of villein labour each autumn.

It is recorded that in 1318 the village had 28 houses. An earlier, undated, document mentions 28 crofts and 26 adjacent tofts (houses) so the village of Wilstrop had at least 26 separate households at that time. A chapel is also mentioned at this date, and two field names—which no longer exist—South Field and Mill Field.

In the poll tax of 1377, 33 persons were taxed, and in 1379 the names of 25 people appeared on the list. Two of them were fullers, so it seems that the mill was probably used for fulling. It is said that boats sailed up the River Ouse and a short distance up the Nidd, possibly to visit the mill.

About 1490, the Wilstrop family—who took their name from the village—enclosed the fields and evicted the villagers. The Wilstrops, who seem to have been a most unpopular family, apparently carried on a running feud with most of the neighbouring gentry. After their eviction the villagers joined forces with some of the gentry, including Sir William Gascoigne and the Abbot of Fountains Abbey, and together they attacked the new park made by the Wilstrops out of the village common.

On one occasion a crowd of some 400 persons came by night and pulled down 1,000 yards of the new paling round the park. Sir Myles Wilstrop had the damage repaired, but it was not long before a party of 200 people came and they: 'pulled down the said pale, and therwith not being content cruelly cutt all the said pale bordez in sonder, and also hewid and kit donn 100 walnottreis (walnut trees) and appiltreis grafted ii or iii yere before, and also distroid the same length of quickwood where the same pale stood.'

Later they found the Wilstrop rabbit warren and destroyed the burrows. They breached the millpond on the same visit, and finally 100 men, tenants of Sir William Gascoigne, hunted Wilstrop's deer and took away all the venison. They also sought Wilstrop in his own house, saying that they would 'repare agayn to the said park and dwelling place of ye said Myles; and make search for him, and if they myght fynde him to sle hym.'

The victims of the many disputes which took place at this time were not always the Wilstrops, as this story—told by Simon Robynson, parson of Moor Monkton, in 1514—will show. In a petition to the King in the Court of the Star Chamber the Rev. Robynson names the depopulator of the village as Sir Guy Wilstrop. He accuses Guy of destroying his tithe corn as it stood in the fields ready for collection. Guy also prevented the parson's sheep from pasturing on the commons of Wilstrop, stole tithe corn from Moor Monkton, and stole Robynson's building timber and some of his animals. On top of all this he fined the parson in his manor court and vexed him with lawsuits.

But worse was to come. On Sunday morning Guy encouraged a man called Joseph Ughtred to assault the parson in his own church during morning service. It is not surprising that, having gone to the Star Chamber, Simon Robynson accused the Wilstrops in the following words: 'in defraude his father, Milles Wilstrop, and he dydd caste doune the town of Wilstrop, destroyed the corne feldes and made a pasture of theym, and hath closed in the common and made a parke of hytt.'

The site of the village, depopulated by the squire of Wilstrop so long ago in order to create more profitable sheep and cattle pastures, can be clearly seen from the air. It lies north-east of the Hall, between us and the railway, on both sides of the present farm-road. Buried here are the foundations of some 26 farmhouses and cottages.

On 4th February 1536, at the first Dissolution of the Monasteries, the prioress of Nun Monkton surrendered the prior and all its lands to the rapacious Henry VIII. Sir

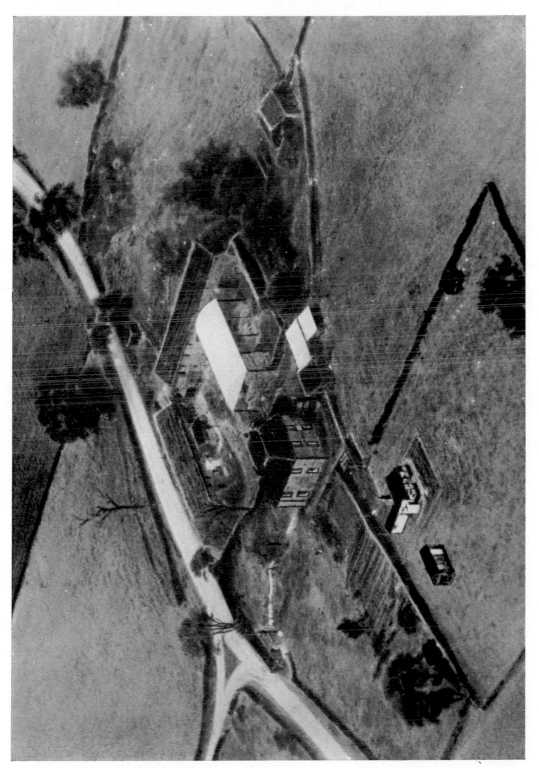

9. Hunshelf Hall

10. Wilstrop Hall, Kirk Hammerton

Oswold Wilstrop asserted that the prioress had no right to certain lands in Wilstrop. He was required to prove this before Whitsuntide 1539, or the land would be claimed by Lord Latimer. His claim failed and the land was added to Latimer's estate. At the dissolution the rent paid to the priory from the Wilstrop land was £2 13s. 4d. per annum, worth about £250–£300 today.

I have been unable to fill the gap between 1539 and 2nd July 1644, when the forces of two opposing armies gathered on the moor near Wilstrop Wood. It must have been a stirring sight when Prince Rupert and his men, with banners of many colours, met the might of Oliver Cromwell and his Roundheads.

The main battle took place to the south of the wood, on Marston Moor, but the defeated Royalists galloped along the woodside, pursued by their enemies, and the following morning the wood 'was white with the bodies of the slain, and the stream in the wood ran red with their blood'. Prince Rupert, it is said, hid in a beanfield beside the wood and his dog, Boy, was killed in the area. It is interesting to note that until a few years ago beans still grew in a field near the wood.

A few thatched cottages stood in the fields at the time of the battle and legend has it that a girl ran from one of them to open a gate for the fleeing Cavaliers. In their haste and fear they galloped over her and swept on, through the yard of Wilstrop Hall and away down the road to York, leaving her battered body in the lane.

After Marston Moor, when local people buried some 4,000 bodies in the wood and the fields around, Wilstrop seems to have entered a more peaceful era, and from the beginning of the eighteenth century I am, at last, able to separate the Hall from the rest of the estate. A family named Gray were tenants of Wilstrop Hall for over a hundred years. It all started when Ambrose Gray married Mary Fawcett, of Wilstrop, on 11th May 1731, at Kirk Hammerton. For many years the farm passed from father to son, no less than four Ambrose Grays being mentioned in the story. Wilstrop has always provided one of the churchwardens at Kirk Hammerton Parish Church, and two Ambrose Grays held office between 1755 and 1777. A list of churchwardens has been kept and the first name on the list is that of Henry Spink, in 1754; the present one is my husband, Wilfred Blacker.

To return to the Grays; an Ambrose Gray was tenant, and Andrew Montague was the owner, on 1st April 1840, when the Apportionment of Rent Charges, in lieu of tithes, took place. Field No. 18, on a map dated 1841, was named as Stoned Horse Close (I didn't know that until I saw the map, today it is known as Stony Field) and had been subject to tithes. From April 1840 it was exempt by payment of a modus of £8 to the perpetual curate of Kirk Hammerton.

In 1870 big changes took place at Wilstrop Hall. The Montagues still owned it, but the tenant was now John Harrison, a man who left many reminders of his tenancy on the farm, and in the surrounding district. A plaque on one end of the present farmhouse tells how the South—or Garden—wing was rebuilt in that year. From an old book I learned that the house which was demolished was 'formerly of stone of very ancient

G

date'. This I believe, as many beautiful pieces of stone are still to be found in the garden.

'The walls of Wilstrop old hall,' the book continues, 'were in parts two yards thick, and during the process of demolition were found human skeletons, antlers of red deer, old coins, etc.' From another book I found that John Harrison had many relics from a bygone age, including some ancient mill stones, called querns. These, I felt, must still be around, and a search found one embedded in the floor of a cowshed, and another in a pathway round the old foldyard.

John Harrison, a devout man, was churchwarden for Wilstrop at the time of the re-building of part of Kirk Hammerton Church. A great benefactor to the church, he died before the alterations were complete, his loss 'bringing a deep grief into the general joy'. He died in 1890 and in 1896, on Palm Sunday, a beautiful stained-glass window given by his sister in his memory, was dedicated. It is in the Lady Chapel but bears no inscription nor anything to connect it with Mr. Harrison.

On the farm he left other reminders. Part of Wilstrop Wood was replanted in his time, and he planted many unusual trees on the river banks and in the hedgerows. A keen ornithologist, the buildings put up in his time were fashioned with spaces in the walls for nests, they are still used today by many birds and one of the buildings is known as 'Starling Castle'.

Now we come to the twentieth century, and events still in living memory. Since John Harrison's day the farm has changed owners twice and tenants five times. During the time of the Egertons, about 50 years ago, the Hall had a staff of servants, including—so I'm told by old people in the district—a butler, cook, and grooms. The butler would be horrified if he could see his pantry today; it is used as a store and the servants' hall houses an electric mixer for mixing calf milk, veterinary supplies and a motley assort-ment of coats, rubber boots, etc. A relic of these more spacious days is a row of bells, hanging on the kitchen wall. When we came, ornate bell pulls hung in all the main rooms, both upstairs and down, but they were such a temptation to our three small children that I had them removed. We still ring the old bells on high days and holidays by means of a long-handled sweeping brush!

Outside, too, are traces of this time. A drive, north of the house, coming through the orchard and along the riverside, was planned. For some reason it was never finished. At other end of the proposed drive shabby ornamental gates still stand and a row of laburnum trees, alternating with red may, marks the route of the road that never came.

The Winter family farmed here until 1944, when we took over as tenants. This time Mrs. Winter was churchwarden, and her husband was noted for the splendid droves of some 50 Lincoln Red bullocks, which he sent from Wilstrop at regular intervals, walking them from Wilstrop to Wetherby Market, 8 or 10 miles away.

In 1949 the estate was sold again this time to Messrs. Sherlin Property, who are still the owners. At the time of the sale the rent of Wilstrop Hall was £450, or £1 4s. 7d. per acre! During the next 10 years several major alterations took place. Mains water, electricity, and cattle grids were great improvements and later a big covered yard was added to the buildings. At the same time half of the Dutch barn, which was unsafe, was demolished and rebuilt.

We discovered a cobbled pathway round the old buildings, and the walls of a small cobbled courtyard, near the back door of the house, shows signs of other small buildings. Two of them, in fact, were demolished after we came. In the farmyard, and curving towards the river, one can still see the remains of a moat, and we have found several wells round the house.

The farmhouse, today, is big and rambling, with no fewer than twenty-one rooms—if we include attics and cellars. As well as the south wing, already mentioned, there is a much older wing, facing north and overlooking the river. The rooms at this side are low, some have oak beams and shutters and some, I think, the original fireplaces.

The rooms on the south side are large and lofty, with marble fireplaces in drawing-room, dining-room, morning-room, and two bedrooms. Two large bedrooms, reached by a separate staircase from the kitchen, were used for staff, in the days when men and maids 'lived in'. More recently they have been playrooms for children, and 'dens' for teenagers. Today the old house is quiet and empty, my husband, younger son and myself being the only inhabitants.

LOST VILLAGE

Beneath this turf, and furrows ploughed with pride,
The relics of a long-lost village hide.
Her peasants tilled this land, now in our care,
Fished in the river, hunted fox and hare
In Wilstrop Wood and over Marston Moor;
Our cattle graze where theirs fed, long before.

Here, in this lonely field, they spent their lives,
Grew food and hunted; here their sons and wives
Picked nuts and berries. Were the woods as gay
With broom and bluebells as they are today?
Gone is their village—vanished—lost—and we
Still farm this land in continuity.

Fourteen Generations in Teesdale

Fourteen Generations on the Same Farm

The outstanding feature of Mrs. Mollie Dent's article on High Green at Mickleton in County Durham is the fact that no fewer than fourteen generations of Dents have farmed there without a break from the 1520s to the present day. Their family records, especially their account books, make a fascinating story of farming in past centuries. One wonders how much farther back they could be traced at High Green if the records were available. Presumably they originated and took their name from the village of Dent, not many miles away on the other side of the Pennines, but when the first of this name crossed over to Teesdale is not yet known. The pedigrees of yeoman families are extremely difficult to trace before the 1520s, mainly because of a lack of records giving surnames. The fifteenth century is particularly barren of records for this purpose.

Marvellous though the Dent record is, it is (unless Mrs. Dent turns up with something else later on) exceeded for longevity by two farming families in the wilder parts of Devon. The Seccombes take their name from their farm in the parish of Germansweek in west Devon, and are first recorded there in a deed dated 1310. They are still farming at Seccombe. Whether they go back farther than this, we do not know. Again we run into a lack of records.

The Reddaways who farm at Reddaway in the parish of Sampford Courtenay, on the northern edge of Dartmoor, are recorded there as far back as the year 1238—more than 700 years in the same place. The Reddaways have spread around the district over this great stretch of time, and farm in several places round about. Their story, if it could be compiled, would make some real English history—not the weary stuff about Top People—and would transform the teaching of history in schools. The Dents of High Green are in this class, too, and one wishes Mrs. Dent had had more space to quote from the records of this old yeoman family.

*

HIGH GREEN FARM
by Mollie Dent

High Green farm is situated at the east end of a village called Mickleton on the Yorkshire side of the River Tees. A road from Barnard Castle to Middleton-in-Teesdale forms the main street of the village and passes the farm gates.

The land rises from 750 ft. to 1,000 ft. in the hills of the Pennine Range. We farm 67 acres, 13 of these are rented, about 24 are in meadowland, the rest rough grazing.

At the time of the survey for the Domesday Book, 1086, Mickleton was described as a waste about a mile long by half a mile wide. This measurement would now approximate to the size of the village.

We can trace at least fourteen successive generations of Dents who have held land in the township of Mickleton, beginning with Nicholas Dent who was on Mickleton Manor Jury in the 1520s; and his son Michael appears in the Mickleton Manor court rolls from 1562–74.

After that Dent succeeded Dent at Mickleton, and we have many of their family records—wills, leases, inventories, and, in a way most interesting of all, the account book of John Dent who lived from 1728 to 1815. He succeeded to the property in 1747, and seems to have had a natural flair for business as the rest of his long life was spent renting, buying and exchanging ridges and parcels of land, and piecing it together into small farmsteads for his increasing family.

By the time their third child was born he built the new house in which we now live with 'John and Mary Dent 1752' carved in stone above the front door. It is an attractive building to the south, built of dressed stone with grey slated roof and nine stone-jambed windows looking over the garden and farmland to the Bail Hill plantation on the 1,000-ft. boundary on the horizon.

We have an account for the stone-work which amounts to £6 0s. 7d. and includes the yardage for the Fore Side, East End, Back Side, Staircase, Sharp, with Hewn Stone for water tabling, Fire Heads, James, Corbales, Windows, Riggan, Chimneys and Stairs, and 11s. 8d. for Dags Working (inscription).

There are the remains of an older building at the west end of the house but all is under one roof. The floor at this end is still cobbled, until recently was without windows and the woodwork shows it has been used for stabling but is now our coal house and meal-store. Above there are two floors each divided into two rooms, with a south-facing window of twelve 6-in. panes of glass on each floor and a north-facing dove-cote.

This uninhabited part is entered from the farmyard by an outside flight of stone steps and is used now as joiners shop and general lumber room. Among the collection of hay-rakes, hen-crates, wire-netting, etc., there is an old sand-horn for whetting or sharpening scythes and a blade with a handle about 2 yds. long for cutting turf.

The back of the house has a much larger expanse of stonework as it is built into a slope and has only five windows, much higher than the ground. Stone steps lead up to the back door which overlooks the road with a view across the Tees to the Eggleston and Middleton hills. This is the official entrance to the house entering what was at one time the back kitchen but now a cloakroom with boiler in place of the old black-leaded range. Originally steps led from here into a cellar but this was blocked up towards the end of last century. The dairy is opposite with north window and a stone slabbed table over cupboards for storage, and shelves above where butter was once kept. There are also wall cupboards, a hanging meat safe fitted with hooks for meat and game, and, until we sent them to the Bowes Museum, a coffee and pepper grinder were fitted to the cupboard top. In the centre of this floor, before it was cemented and painted, there were hinged wooden doors over a pit used for storing potatoes. The original stone stairs went up above from this part of the house. A door was broken through the wall into the old building in 1922 so that the milk could be brought in to be separated without tramping through the rest of the house. The pigs for bacon were always salted down in a lead bowl in here before being washed and hung to dry in the kitchen: the cream matured in earthenware pots before being churned in the large end-over-end churn and then worked into butter. Now it is more or less a utility room where the electric washer and spin dryer are stored.

At the front of the house, on the ground floor, the sitting-room, which in turn has been the dining-room and the parlour, is the only room with a boarded floor. The others, until 1922, were all stone flagged; these have since been replaced with coloured cement. About twelve years ago, when taking out the white Italian marble mantelpiece to replace it with a tiled fire, the builder uncovered an arch of red bricks in the wall. This was obviously an earlier hearth but looked small for the size of the present room.

The front kitchen is to the west of the centre passage and, like the other room and bedrooms above, has two windows. At some date a small closed-bed has been added here, built of brick and jutting into the old stable area. It is hidden behind panelled doors; this panelling continues the full length of this wall and at one time ran halfway along the next, as a delft-rack filled with willow-pattern plates and dishes, with drawers and cupboards under. The moulding round the wall has been interrupted to take a grandfather clock. This clock was bought in 1839 when William Dent and Mary Parkin were married. It is made of oak with veneering, was unpolished but she gradually burnished it by rubbing with beeswax; it is still wound daily and keeps good time. The face is decorated with a painted scene of bird shooting, the maker—Thwaites, Barnard Castle. In this room, the high grate with hobbs at each side was replaced in 1947 with a green enamel combination range, the stone mantle being left in position. At this time the green flooring was laid and the closed bed turned into a pantry, a sink and fitments were put in and the room became living-dining-room cum kitchen. This year a re-frigerator has been added.

John Dent's account book (already referred to) contains a wonderful amount of information about prices, wages, and so forth. Here are a few extracts out of many I could select:

In 1765 his outgoings included

	£	s.	d.
Lord's Rent	3	4	4½
Land sess (tax)	2	10	4
Window sess		14	0
Poor sess		16	0

	£	s.	d.
Church sess		6	3
Constable sess		6	6
Tythe Rent		19	4
Legacy left by an Uncle to poor		5	0

In 1792 Tythe to the Rector:

	£	s.	d.
2 lambs @ 3/6		7	0
5 fleeces @ 1/–		5	0
5 Communicants			7½
1 Swarm of bees			1
1 Foal		1	0
½ a calf		13	0
4 cows @ 2d. each			8
1 Farrow cow			1
House			2

In 1755 he bought:

	£	s.	d.
A Bacon Flich 2 st. 4lb. @ 5s. 3d.		12	0
½ share leg of veal, won at cards, Grassholme			6
½ lb. lofe sugar			5
2 bushels of corne		3	7½
10½ lbs. cheese @ 2½d.		2	2½
Quarter mutton		1	2
14 st. of beef @ 1/8d.	1	3	4
6 pecks of barley		3	0
A milk cow	4	17	6
Joseph Rane's mare	5	15	0
9 yds. of lining cloth @ 1/11d.		17	3
Stockings		2	2

In 1771 he was selling:

Feb   17 lb. Butter @ 9½d.—weekly
June 12 lbs. Butter @ 6d., first new milk cheese made
Oct.   26 lbs. Butter @ 9d., Dec. 29 lbs. @ 10d.
Sept. 21½ lbs. Cheese @ 3½d.
      Flitches of Bacon @ 6s. a stone
      Hay @ 8½d. per stone, £ 8 8s. 0d. for a stack of hay

3 Thraves of Rye straw 3/–
1 Thrave of Barley straw 6d.
Pigg Heads 4d., Hams 5d. per lb.
Winter eatage Back of the Hill to Jn. Foster 1/6d
Thos. Kidd Cow 18 days Fogg 10/–.
Others 3/–, 3/6, and 4/– a week.
Drying oats for piggs 10d.
Keep for a quy 2/– per week
12 Bushels of barley bought 7/–

Still 1771—Paid out.

4 pecks of salt 5/–
2 Bushals of Masilgin for seed 12/1
1 Bushal of wheat 7/4
Opening gutters 6d.
2 days hedging 1/–
For spinning 6 lbs. of wool for mother 2/6
Ale for a cow 6d
3 Beasts £16 10s. 0d.
1 peck of apples 3/6d. Shearers 9d.
Setting tatos 1/6d. Taty gatherers 1/–
Tatos set at Longlands 22 pecks
Tatos set at two Head Rigs in the Middle Field 10 pecks
1780   4½ days mowing 4/6, 2 cows bulling 2/–.
1781   2 quarts of tar @ 8d, Ploughing 3/–
1789   22 wether lambs 6/– each, 9 gimmers @ 6/3 each
         11½ st. of wood @ 7/– £4 8s. 6d
1794   85 sheep salved

The year 1802 saw the Enclosure Act brought into action when the Town Fields, the East and West Pastures, and the Moors were divided up among the landowners, part being sold to help defray the expenses.

Of John Dent's claim of leasehold land 27 acres 3 roods and 10 perches were in 54 'parcels'; and in the 24 acres, 3 roods 36 perches of freehold claim there were 64 'parcels'. This enclosure set in motion a widespread demand for stone walls and fencing. A bye-law said that in the new inclosure no sheep or asses were to be kept for ten years to allow quicksets or newly planted fences to grow. His share of the expenses for division and fences, timber, crops, £446 5s. 9d.

A bill presented by Wm. Raine and Partners in May 1804 was for 137 roods of walling in the Brigstones, Longlands, Skellabushes, Rowberries, Hartelknots and East Row-berries: @ 8/6 and 9/– per rood it came to £59 10s. 2d. The wall between the Skella

Bushes and Pasture Banks was built in a wavy line to allow for shelter on this north-facing slope.

Another major upheaval in the history of the farm was the making of the railway from Barnard Castle to Middleton-in-Teesdale which crossed the upper fields of Hartel-knots, Rowenberries and Longlands, meaning redistribution of field boundaries and watering places, as also the sale of $1\frac{1}{2}$ acres to the Railway Company in 1866.

In 1937, William Dent, then a widower, was pleased to turn over the responsibility of the farm to Thomas, his only child, soon after his marriage.

During the thirty years we have farmed here we have gone through a farming cycle. Starting with butter and cheese, gradually building up a milking herd, altering and having erected new buildings so that we could milk a dozen cows, then eventually having to cross our Northern Dairy Shorthorns with Friesians to meet the fashionable demand for black and white cattle. This spring we sent our last consignment of milk to the dairy and are now concentrating on rearing with suckler cows taking the place of milkers.

Our herd now numbers 40 to 50 head of cattle. Of the young stock the black and white heifers will be reared and sold newly calven, the bullocks and store animals sent to the Fatstock Marketing Corporation when finished fat.

With a flock of 35 Swaledale ewes crossed with a Teeswater ram (this year a lamb crop of 182 per cent) to produce the Masham lambs, all lambs not sold for breeding are also fattened and go to the Fatstock Marketing Corporation.

A small flock of Pedigree Teeswater ewes are kept, shearling rams and surplus ram lambs being in good demand.

At one time we kept about 150 hens but now the deep litter house is used for fattening the produce of two sows.

During the war years we ploughed a little out for roots and oats, the latter being cut green and dried for hay as the growing season here is too short for ripening grain. Over the years we have grown good crops of kale for the milkers and when re-seeding was done a cover crop of rape was sown for fattening the lambs.

For a number of years we have made silage as well as hay but this year (1968), not having a dairy herd, we will need more grass for grazing young stock; also as the cattle are wintered inside from November to May it is more convenient to feed hay rather than cut and carry silage.

At one time bracken and rushes, even hay seeds, were used for bedding, but of late we have bought straw from farms lower down the dale, either leading the bales with tractor and trailer or hiring a wagon.

We also have a very productive fruit and vegetable garden, growing most of our own supplies.

What will happen to High Green in the future? Small farms are supposed to be uneconomic and many would like to see them go. Our family consists of three married daughters. Of course there are grandchildren coming along.

Three Scottish Farms

Three Scottish Farms

The three farms that follow are all in Scotland—one in the far north, on the Caithness coast; one on the bleak northern coast of Aberdeenshire; and one in the uplands of southern Scotland, on the borders of Dumfriesshire and Kirkcudbrightshire. Though they are all very different today, they all have one thing in common and that is the impossibility of writing anything like a continuous history as one can for so many farms in England. In some parts of England one can write a history beginning with Domesday Book—and sometimes from well before that—and every century produces more or less useful documents to help on the story. But Scotland has had such a turbulent history that it has succeeded in destroying almost all its local records before about 1700. Craiglearan in Dumfriesshire is fortunate in having a solitary reference as early as 1507, but for most of Scotland the historical picture, when it comes down to villages, hamlets, and farms, is a blank for several centuries. It is only with the establishment of more settled conditions after the Union with England in 1707 that local records begin to accumulate and survive.

On the other hand, two of the three farms described in the following pages are rich archaeologically. There is Summerbank in Caithness, with its magnificent *broch*, showing that farming and fishing were carried on here some 2,000 years ago; and at Craiglearan in Dumfries there is a whole series of archaeological remains dating from the Bronze Age onwards, showing some 3,000 years of farming on these hill-pastures. But until the rich archaeology of Scotland is properly explored by experts, and much more excavation done, we shall know very little of early farming in these northern parts. And after the archaeological record ceases, there is an enormous gap until the documents begin. We shall never be able to write the story of farming in Scotland as we can write it for nearly everywhere in England.

Nevertheless, there are some compensations. Farming history does not have to be ancient to be interesting, and the account of the last 80 years at Pittendrum on the Aberdeenshire coast shows how fascinating recent history can be. After all, 80 years— or even 40 years—ago, is another age, and the changes that have taken place on the farm in the last generation or two deserve to be recorded as faithfully as those of earlier centuries. Farming and farmhouses have undergone a revolution within living memory, and it is important to record it in detail while the changes are fresh in the mind. If only farmers and their wives had thought of doing this in say the Tudor or the Stuart period, what a record we should now have of the lives and the houses of ordinary people and of the domestic changes they witnessed during their obscure and useful lives!

111

THREE SCOTTISH FARMS

CRAIGLEARAN, DUMFRIESSHIRE
by A. B. Hall

Craiglearan is a hill farm of 1,186 acres in the upland country bordering Kirkcud-brightshire. Indeed the farthest boundary of the farm marches along the county boundary at a height of some 1,100 to 1,700 ft. above sea-level. The steading lies at the south-east corner of the farm, where the burn enters the lowlands, about 700 ft. above the sea, but the hill-gazing rises to 1,700 ft. at Wether Hill. The nearest settlement of any size is Moniaive, about 4 miles away.

The name Craiglearan is Gaelic in origin. The *crag* is Craiglearan Crag on the frontier of Dumfriesshire and Kirkcudbright, and the second half of the name seems to come from *leargan* or *leargain*, which means 'a sloping hill or a steep pasture-ground'.

Though the house only dates from 1880, it occupies an old site; and there are at least six sites on Craiglearan showing signs of habitation at different times in the past—all awaiting professional examination or excavation. It was not until 1967 that they were visited by anyone with any archaeological knowledge, when three of the sites were tentatively estimated to date from the Bronze Age.

Between the dipper and the first plantation are five or six 'platforms' or level areas scooped out of the hillside, which could have been the sites of wattle or timber/hide houses. Richard Feachem in his *Guide to Prehistoric Scotland* describes an unenclosed platform-settlement in Peebleshire dating from perhaps the fourth century B.C., which could help to date the settlement on our own hillside. There are several stone heaps near by which may be burial cairns.

A little to the east is a circular 'house' with an outside diameter of some 19 ft. This type of house is dated by Feachem as not earlier than the first century A.D.

On the knowes (knolls) below the Craig there are more circular buildings and cairns, and by the Carrock bield more buildings, or rather their foundations. There are several well-defined rectangular buildings in a semi-defensive position on a narrow ridge between Craiglearan burn and Jerkney burn. One of the largest is 15 ft. 8 in × 12 ft. inside; and a small one is 8 ft. 6 in. square. Two upright stones (one 2 ft. high, the other 1 ft. 8 in.), standing 6 ft. apart, look very much like the gateway of a perimeter wall.

The two other main sites are probably those shown on a map made in the early seventeenth century. Again, there are some obvious buildings in the Craig Park surrounded by patches of crop marks. By the Carrock bield are some very tumbled-down stone walls in one of the most rock-strewn areas of the farm.

But after this wealth of archaeological remains, there is a long silence in the history of the farm. The first documentary reference occurs in 1507, when King James IV of Scotland granted a charter to Robert, lord Crichton of Sanquhar and his heirs, of

112

11. High Green, Mickleton

12. CRAIGLEARAN

13. PITTENDRUM

14. Summerbank Farm: the broch

various lands which included 'two and a half merkland of Craglereane, with Fyglane', that is, land to the value of 2½ marks or 33s. 4d. annually.

On a map drawn by Timothy Pont about 1595–96 there seem to be three farms on Craiglearan, known as Kraiglenan, Over Kraigleiran, and Kraigtom Hill. In 1671 the farm was part of the Maxwelton estate, owned by the Laurie family. A so-called 'cave' on the farm, consisting of a large stone slab round which rocks have been built, leaving a small entrance, has been described as a 'Covenanter's Cave'. This, if true, would date its occupation to the period 1662–85 when the Covenanters were being hunted out of the parish by redcoated dragoons and local sympathizers including the laird, Robert Laurie.

On the famous map of General Roy, who has been called the father of the ordnance survey, as he set an example for accuracy and detail, showing woodland and field boundaries among other features, no boundaries are shown at Craiglearan. This suggests that in *c.* 1750, when the map was made, the farm was still largely unenclosed hill pasture.

We hear of the farm again in an Edinburgh newspaper advertisement dated 21st November 1786, announcing the intended sale of parts of the Maxwelton estates. *Craiglyrian* is then described as of about 740 acres, whereof only 17 were arable. With it went the lands of Meikle and Little Laggan, consisting of about 284 acres, whereof 69 were arable, 9 were meadow-ground, and the remainder good pasture land. The two farms were let together for £121 18s. a year at that date.

A valuation dated 1827 shows that Robert Kirk of Craiglirian then owned the farm, and he was still owner in 1835 when a splendid hand-coloured plan of the farm was drawn, showing field-names and giving acreages (Scots statute measure) totalling 931 acres. The layout of the fields and woods was by this time very similar to what it is today, and the Ordnance Survey map of 1856 confirms the details. Of the 931 acres in 1835, only 17 were arable (just as in 1786), 13 were meadow ground, 5 were woodland, and no less than 896 acres were pasture.

In 1879 Craiglieran was bought by James Hewetson, who built the present house—date stone 1880— and the main range of buildings. After various further changes, Hugh B. Hall took over the sheep stock after the bad winter of 1947, and came to live here in the October of that year. The land has subsequently been farmed by his wife and son.

There are two main sets of sheep-handling bughts, both of drystone construction. The top hill sheep are dealt with in the Craig Park at bughts which are perhaps a relic of the old farm there. A similar layout near the steading is adequate for the bottom hill handlings and clipping, which is done by hand. A *bought* or *bught* is a Scots word for a sheep-pen or fold.

Though farming has been carried on at Craiglearan for perhaps 2,000 years, or even 3,000 if we think of the Bronze Age sites, very little is known about it. It is fairly safe to assume that it has always been hill-farming either for sheep or cattle, with small

H

patches of arable near the farmsteads. Before the late eighteenth century there were two, or perhaps three, farms with some small fields enclosed by stone walls. The remains of their cropping areas are still visible on the steep, dry, rock-strewn hillsides. Hay, potatoes, and oats were probably the chief crops. The latter, prior to grinding into oatmeal, may have been taken to a neighbouring farm where *Kiln Knowe* is shown on the 6-in. map. This was a 'killogg' used for corn-drying. Rearing black cattle for sale in England is said to have been the main enterprise then.

Sheep, probably Blackface, took over from cattle and the boundary walls (what we call 'march dykes') were built to keep them under some sort of control. Many miles of hand-dug drains, parallel to each other and running straight downhill, were dug during the last century. Modern ploughed drains cross the slope. The sheep stock must have changed to Cheviot at one time, and the Bottom Hill at Craiglearan was known as the Cheviot Hill until quite recently. Later, Blackfaces were re-introduced, and, as at present, bred to Blackface tups. The Top Hill ewes are hefted into four—Meadow Head, Wether Hill, Burnhead, and the Craig.

In recent years the stock has been 520–530 ewes. Ewe hoggs, which number about 130, are still wintered away in Ayrshire, and until 1947 were driven over the hill to New Cumnock, a journey of some 15 miles, which took two days every October and April.

The present herd of 18 hill cows of predominantly Galloway blood has been built up over the last few years, after an interval without outwintered cattle. Dairying, heifer rearing, bullock fattening, pigs and poultry, have at different times been tried, but all on a very minor scale. Most of the ploughable in-bye land has been cropped and re-seeded since the war, but hay is now the only crop attempted. Ten to twelve acres are about as much as can be mown.

*

PITTENDRUM, ABERDEENSHIRE
by Christian Milne

Pittendrum, Sandhaven, Fraserburgh, Aberdeenshire, is a 90-acre arable farm lying in the fertile, coastal belt along the exposed, rock-bound coast of the Moray Firth. There I was born, and there my father farmed for 57 years, his uncle having farmed there before him. The farm was rented from the Pitsligo estate which was then entailed.

The farm was extremely fortunate as regards water. The Lady's Well, from which the farm water supply was drawn, was of excellent quality, a spring babbling up in limitless abundance from a bed of silver sand. There was also a large dam at either end of the farm. A lade* carried the water to drive the wheel which powered the threshing mill in the barn, another served the miller with power for his meal mill. There were, accordingly,

---

* *Lade* is an old English word for 'a drain, a water-course'. In South-West England it is still used in the form *leat*, especially of a mill-leat.

many natural watering places for the cattle to drink at, as well as abundance for all other farm purposes.

When my father went there in 1887, and for many years afterwards, the farmhouse supply came from a dip well at the end of the avenue, and I can just remember when every drop needed for domestic purposes was carried from it—in pails slung from a wooden shoulder yoke. Later the water was piped, coming by gravitation from the Lady's Well to the cottars' supply, the horse trough and the farmhouse. Many years later, modern conveniencies were installed in the farmhouse, the water for these again being provided by the above-mentioned supply.

A field was called after the Lady's Well, another was called the Eppie Field—after a

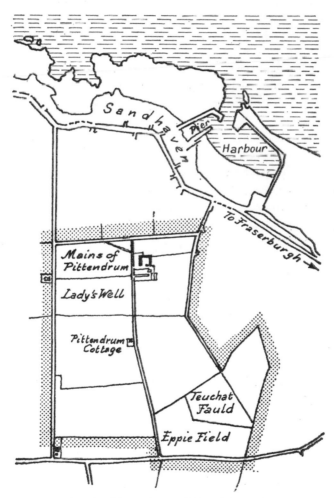

FIG. 9. Pittendrum (Aberdeenshire)

long-ago ancestress of mine who had a croft there. But perhaps the most interesting was the Teuchat Fauld, for here the lapwings gathered in their thousands when a storm was imminent, finding its low-lying position a sanctuary from the terrific winter gales which swept in, unobstructed, from the Arctic.

In Pratt's *Buchan*, published in 1859—the standard work of reference for this area—we read: 'We pass on to the farm of Pittendrum on the "muir" of the same name.' Pittendrum was originally part of a small estate, and in the old parish records, the adjoining village of Sandhaven—5 minutes' walking distance away—was known as The Port of Pittendrum. It was then the ancestral home of the Cumines of Rattray, a family who were robust Jacobites and staunch supporters of Prince Charles Edward. Their coat of arms—a rose, a thistle and a shamrock in carved relief—still adorns the arch above the farmhouse front door.

Pittendrum was built in 1734. It is a stone-built, three-storey house, with 4-ft. thick walls. These are essential to withstand the salt-laden gales, for it stands—open and unsheltered—on the most easterly point of the exposed north-east coast of Scotland. The old fishermen have told me how Pittendrum's tall shape was visible far out to sea, and was used as a landmark by the fishing boats. It consists of eight rooms with kitchen, bathroom and cellars.

When my father went there in 1887, the house was gutted and renovated. Before that most of the rooms were oak-panelled, with stout wooden shutters on every window, and in place of large rooms there were several small ones with dressing-rooms attached. The staircase was then on the eastern side of the house, instead of being on the western as it is now. During the renovation a curious fact emerged, for, to insulate the rooms and also to deafen them, straw ropes—made with a wire straw-hook—had been used. I remember my father telling me there were cart loads of these straw ropes to dispose of when the house was gutted.

The kitchen—fortunately—was left untouched, for I never remember seeing a more beautiful roof than the deep, gleaming, satin-smooth, rich brown rafters. In passing I may explain that peat reek—smoke to the uninitiated—and time had given it that wonderful mellow glow.

I can just remember the low, peat fire in the kitchen. It had a stone deece—bench to the Sassenach—on either side, and took a whole barrow load of firm, black peats to re-fuel. It was 'rested' at night with damp peats and smored with a depth of peat ash which never allowed it to go out, a puff with the bellows being all that was needed to bring it into glowing life for the 5 o'clock rising-time.

It may be of interest to mention that in each lease on the Pitsligo estate the tenant had a right to a lair in the peat moss situated some 12 miles away. The peats were cut and set by the farm staff and left to dry. Later they were carted home and built into stacks.

There was also a huge, built-in girdle—hot plate in modern parlance—separately fuelled and connected to the kitchen chimney by a damper. On specially busy days such as steam-mill and potato-lifting, the girdle came off and a huge iron ring—made at the local smithy—was substituted. This ring had an open circle in the centre, and the speed with which it brought a large pot of soup to the boil was a record. The kitchen walls were white-washed and the floor stone-flagged.

In the milk and coal cellars the stout, unpolished wooden rafters had been hand-chiselled from the solid trees, and these, too, in 1968, are still in good condition. The house has 26 windows, also five blind ones, doubtless bricked up to evade paying the tax, which—until 1851—was levied on windows of dwelling-houses.

The house and steading, must, at one time, have been situated in the centre of the farm, but in 1903, part of the farm was given up as fen land for the village, and to make the adjoining meal mill into a croft. This meant that the boundaries no longer extended down to the foreshore. The meal mill was at one end of the farm, the smithy at the other, extremely convenient for milling the home-grown oats, and for horse-shoeing and repairs to farm implements.

Here I may mention that my father was not mechanically minded, and never owned a tractor. This, of course, meant that the whole cycle of operations which comprised the farming year had to be done with the aid of horses.

The farm stock consisted of feeding cross cattle, all housed and bound. My father also bred Clydesdales, cross sheep and a limited number of pigs. In 1903—when he rented another farm—he founded a small herd of Shorthorns in addition.

Poultry did extremely well on Pittendrum and we had always a large breeding stock. They were kept in colony houses on the free range system. I have heard it said that the high iodine content of the pasture contributed greatly to their stamina and high production. I may say I have lived through the transition of the poultry industry. I can just remember my mother hatching all her replacement stock of chickens, ducks and turkeys, by broody hens. I remember the excitement cause by the arrival of our incubator. It was the very first in the district. Soon others followed, and my father was in much request among new owners to explain their intricacies.

We always hand-milked two or three dairy cows, which supplied both the farmhouse and the cottars. The surplus milk, butter and cheese, also the commercial eggs, were absorbed by the local villagers, all being sold direct from the farmhouse door. Bees also did well at Pittendrum, the old high-walled garden sheltering them from the prevailing winds, and the surrounding pasture providing them with abundance of white clover. Honey found a ready sale locally.

Another perquisite of the farm was the right to drive sand and seaweed from a particular stretch of foreshore. I can remember how gleaming loads of wet seaweed were always spread over the dung courts in summer as a seal to prevent the manure drying out. I can also remember huge loads of seaweed passing my home to an adjoining farm, where it was lavishly applied to the land in place of balanced fertilizer. Perhaps these numerous loads of seaweed were instrumental in making the grassland of these coastal farms capable of carrying so many cattle and sheep?

Here I should explain that all the fields of Pittendrum were enclosed by dry-stone dykes. Among my father's papers we found records that these dykes were built by local dry-stone dykers at 2½d. a running yard. Draining the heavy clay at like period—1887— cost 4½d. a running yard. I consider these substantial dry-stone dykes were certainly one of the assets of the farm, being invaluable as shelter. That this was so was proved by the way in which all farm stock availed themselves of them.

Pittendrum also figures in local weather lore:

> *Fin the rumble comes frae*
> *Pittendrum*
> *The coorse weather's a'tae*
> *come;*
> *Fin the rumble comes frae*
> *Aberdoar,*
> *The coorse weather's a'ower.*

The 'rumble' referred to is, of course, the roar of the breakers crashing on the rock-bound shore.

Buchan, the progressive agricultural area in which Pittendrum is situated, has been described as 'a treeless land where beeves are good'. But it cannot always have been treeless, as once, many years ago, when draining operations were in progress in a mossy hollow, a giant specimen of a black oak was uncovered. It was in the field immediately adjoining the farmhouse, and indicates that at one time a belt of woodland must have stood between it and the sea.

Pittendrum grew excellent crops of roots, hay and cereals. When my father began farming in 1887, as a lad of 18, two old farm workers cut the entire crop with scythes. That was before the advent of the binder. Before the introduction of weed-killing sprays, my father grew potatoes as a cleaning crop, charlock being the weed most difficult to eradicate. I have heard the infestation attributed to the lavish application of seaweed and this might well be, since, further inland, charlock was virtually unknown.

As this yield of potatoes had no market, ground was let in drills to the fisherfolk of the twin villages of Sandhaven and Pitulie. A day was set and the local crier went through both villages, ringing a bell and announcing, in a stentorian voice, the actual date and that planting would begin at *7 a.m. prompt*. The villagers themselves then took over, supplying their own seed and doing the planting. The most popular variety then was the old-fashioned Champion. The farm staff then filled in the drills and did all further cultivation for the season. The crop was lifted by the villagers—including the children—and one of the men, with a horse and box-cart, delivered them.

My father grew a purple-top swede turnip seed which was greatly sought after. I remember how carefully he used to select perfect specimens for seed-growing. The plot where they were planted was in a corner of a field beside the stackyard, and I always associate the glorious golden patch with the humming of thousands of bees, for whom it was a veritable paradise. The turnip was richly-coloured, firm and very sweet.

118

Everyone who came about the farm was fed in the farmhouse. Extra helpers for hay-time, harvest, potato-lifting and threshing-mill days, for all the oats and barley was built in stacks and threshed on the farm. Any tradesmen employed, either inside or out, came in for meals. The actual cooking for these was never a problem, a large solid-fuel cooker coping admirably. Supplies were no problem either, as there was always abundance of home-produced foodstuffs—milk, butter, cheese, eggs, pickles and jams—not to speak of the hams and rolls of bacon—cured by my mother from her milk-fed porker—and the strings of oatmeal puddings, made at the approach of winter, and stored in the meat girnel.

The household chore I hated above all others was cleaning the numerous farmhouse paraffin lamps. Added to these were the lanterns for the byres and stable, which meant cleaning many glass slides as well as lamp glasses. These paraffin lanterns were, to me, a source of constant fear about the steading. I was most thankful when they were replaced by electric light, for the fire risk as well as the convenience.

Pittendrum is now owned, and farmed, by my brother, who bought it from the Pitsligo estate when the entail was broken a few years ago.

*

SUMMERBANK FARM, CAITHNESS
by Alistair Sutherland

Summerbank is situated on the north-east coast of Caithness, the most northerly county on the mainland of Britain. The farm lies some 10 miles from the county town of Wick and just 7 miles from the well-known village of John O'Groats. The North Sea forms the eastern boundary of the farm and the main A 9 road runs through part of it.

The steading marks the division of the parish of Wick and the parish of Canisbay. The name Summerbank is of Norse origin—like all the names in this part of North Scotland—meaning the dividing of the parishes. Although the farm has always been known as Summerbank the farmhouse had been a coaching inn. We do not know exactly how old the house is, though the shell is certainly more than 300 years old. The inn was known as The Half-Way House, because it was built half way between Wick and Huna. The hamlet of Huna lies on the north coast where the ferry-boat left to cross the Pentland Firth for Orkney.

The coach house and stable of the inn still stand today, and are used as part of the steading. There is no record of actual dates when the house was an inn but it was certainly used for this purpose up to 1800. When we came to live here 13 years ago the doors inside the house were still numbered.

Probably the first farmers on this land were the broch people who lived 2,000 years ago. A *broch* is a local name applied to ancient dry-built circular 'castles' in the north of

Scotland of which there is a very good example on the farm. The broch of Nybster stands on top of a high cliff promontory about 60 yds. in length and 40 yds. in width. There has been a wide ditch or moat cutting it off from the land. Inside this ditch is a wall some 10–15 ft. thick. Near the middle of this wall is the entrance passage, 15 ft. in length through the thickness of the wall.

Immediately to the rear of the entrance there is the broch itself which is entered from the opposite or seaward side. It is enclosed by a maze of passages and out-buildings which are scattered over the whole of the promontory. This was excavated in 1896 by my grandfather, John Nicholson, who was a well-known historian and antiquarian.

Some of the contents found in this broch are housed in the National Museum of Antiquities, Queen Street, Edinburgh, among them a small quantity of charred grain which shows that some kind of cultivation must have taken place 2,000 years ago. A number of stone implements and household utensils, belonging to this period are displayed in a small museum on the farm.

According to Mr. Geoffrey Grigson in *The Shell Country Alphabet*: the brochs, which are found all over Scotland, 'were built and lived in between about 100 B.C. and A.D. 100 by farming and fishing people of the Iron Age who found themselves in need of tower homesteads or fortified homesteads—British immigrants who had moved up from the eastern lowlands and Northumberland under pressure from Belgic invaders farther south in Britain. Centuries later their tower-houses were named brochs (from Old Norse *borg*, a fort) by Viking settlers or travellers.'

A mile to the north of the Half-Way House and on a rugged, practically inaccessible rock, stands Bucholie Castle, built by one of the greatest Norse pirates, Sweyn Asleifson. A more gloomy and solitary place to have lived in is hardly possible to imagine. Sweyn spent his winters in this castle stronghold with about eighty of his followers. In 1160, on one of his many plundering expeditions near Dublin in Ireland, he was ambushed and slain.

During the reign of King Robert the Bruce (1306–29), Bucholie Castle became the home of the family of Movat, who held it in their possession until 1661 when it was sold to the Sinclairs.

Most people try to visit John O'Groats sooner or later and all the stories and legends connected with this district are fascinating. A certain John Grot did live in what is known as John O'Groats but where exactly his house was and where he came from are not very clear. My family was connected to the Grot family through marriage. The John Grot stone which is believed to have been above the entrance door of John's house in Duncansbay (now called John O'Groats) is built into the wall of the farmhouse here.

The most popular legend is that John Grot came to Scotland from Holland during the reign of James IV (1488–1513) and arrived in Caithness in 1496 with a letter of introduction to William Sinclair, Earl of Caithness, who lived in Girnigoe Castle. He was granted land in Duncansbay, paying yearly a rent of one penny and three measures of malt. As time passed, John's family of eight sons, grew up and as was the custom of that time, each son received a portion of land. To celebrate their arrival in Scotland, they held a banquet on March 14th each year. The sons could never agree who was to sit at the head of the table so John had built an eight-sided house with an eight-sided table, so that all the family felt equal. The present John O'Groats House Hotel was modelled on the eight-sided house which is supposed to have stood where the flag-pole stands today.

The second story about John Grot is that he was a ferryman between the mainland and Orkney—there is a record of the Grot family being ferrymen between 1549–1715. He is supposed to have charged a 'groat' as the fare.

The third legend is that John was a miller and as he kept the husks from the oats as part payment, he was nicknamed 'groat'. The husks and trash from the oats after the meal was made, were called groats.

The fourth story, which I think is nearest the truth, is that John came from Fife. His father was a Dutch trader who settled in that county. He had three sons, John, Malcolm and Gavin. Malcolm Grot became a lawyer in Edinburgh. Gavin, being in the service of the Stuarts who were on the throne of Scotland at that time, received land in Orkney. John was in the service of the Sinclair family in Caithness and for his loyalty and devotion, received land in Duncansbay. As with most legends, parts of all the stories are probably true for we know that John Grot who was a son of Finlay Grot, received land, a ferryhouse and ferry in November 1549.

The Grot family can be traced from this date until 1st August 1741, when a Malcolm Groat disposed of all his lands in Duncansbay and in the parish of Latheron to pay a debt of £8,000. Mr. William Sinclair, of Freswick, paid this debt and afterwards became the Laird of Duncansbay.

At present Summerbank Farm comprises 205 acres, in two separate parts as you can see from the sketch map. The original farm as I inherited it from my uncle and aunt was the southern 83 acres around Summerbank House. The northern part consists of crofts bought and rented since 1958, plus 85 acres of moorland which I have been reclaiming over the past three years.

The soil on Summerbank is medium clay, with typical blue Caithness flagstone rock not very far from the surface in places. This is the famous Caithness flagstone which formed many pavements in cities throughout the world. On all the other parts of the farm is peat, some of it reclaimed a few years ago. Peat is still used today as the main winter fuel, although in many homes it is slowly being replaced by the more modern electricity and oil. Peat is cut with a special spade about mid-May and allowed to dry on top of the heather. Depending on the weather, it is carted home and stacked about July.

Until about 1800 the way of life in this part of north Scotland varied little from the Norse way of life brought to these shores about A.D. 800. It is of interest to note some details of this ancient Scandinavian life, followed from generation to generation. The people housed their cattle in winter near their own dwelling. There the cattle remained from November until April or May. As the year advanced and the *saeters*—the common pasture lands—began to produce some grass, the herds of black cattle were moved to these; men, women and children—all the people went to live at the *saeters* in turf-built houses.

During the winter months the work was chiefly indoors. The making of clothes,

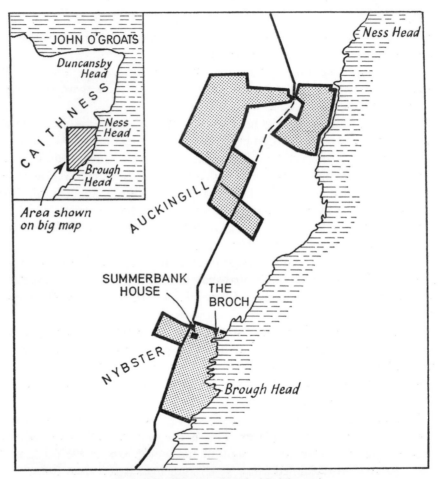

FIG. 10. Summerbank (Caithness)

spinning, knitting, weaving, cooking were the work of the ladies of the household. The oats and bere—crude barley—screws in the cornyard were hand-threshed by the men, the grain was then dressed, winnowed, riddled, sieved and kiln dried. Part of the wall of the kiln can still be seen in the old barn which is used as a workshop today.

The hand-mill was displaced by the mill worked by water-power about 1790, though there is evidence that the hand-mill was used after this date.

THREE SCOTTISH FARMS

With the introduction of turnips into the country in 1790, and potatoes in 1754, the farming way of life changed completely. More cattle were being kept over the winter and the North Country Cheviot sheep were beginning to be introduced into the county by Sir John Sinclair, who was the founder and president of the Board of Agriculture.

When potatoes became part of the everyday diet, combined with the abundant catch of herring, 'tatties and herrin'' became characteristic Caithness fare. Even today we have social evenings with 'tatties and herrin'' as the main course.

Farms and the Industrial Revolution

A South Lancashire Farm

This story of Allwood House in south Lancashire is a fascinating account of a farm which is half old—already a going concern as early as the year 1210—and half the creation of the Industrial Revolution. In this respect it is a complete contrast to the south Staffordshire farm (Brick House, Rowley Regis) described in the last chapter of this book, which was literally undermined by the Industrial Revolution, fell into ruin, and has now been lost almost without a trace.

The 'mosses' of Lancashire were a great feature of its natural landscape. Some were enormous: of one it was said 'Pilling Moss, like God's grace, is boundless'. Chat Moss, which Joan Grundy describes so well, is even better known, particularly in early railway history.

These ill-drained and at times dangerous mosses were not valueless. In summer, when they dried out, they supplied useful common grazing. Chat Moss is first recorded as a name in the thirteenth century; but it takes its name from an Old English landlord, one *Ceatta*, so it has existed from time immemorial. In this respect it is like the great 'waste' of Otmoor in Oxfordshire, which was 'Otta's moor' and was similarly a vast common pasture for all the neighbouring villages.

Actually, Mrs. Grundy is not quite right in saying that before 1800 Chat Moss was an impassable spongy morass. According to Roy Millward's brilliant account of the history of the Lancashire landscape, the Duke of Bridgewater began the process of draining and reclaiming the Moss when he constructed his famous canal, which was opened in 1761. Arthur Young, when he wrote *A Six Months' Tour through the North of England* (1771) speaks of the Duke's work in reclaiming the northern edge of the Moss as one of the finest achievements of the eighteenth century. It was a by-product of the making of the canal. 'The navigation (canal) is carried a mile and a half beyond Worsley, into the middle of a large bog, called here a moss, belonging to the Duke, and merely for the use of draining it and conveying manures to improve it.'

The rest of the Moss was drained as a result of another magnificent engineering achievement—Robert Stephenson's building of the Liverpool and Manchester railway across it, despite all the pessimists who said it would never support a loaded train. The work was begun in June 1826 but for a long time the Moss swallowed up his labours, until he thought of laying a foundation of overlapping hurdles, on which he then laid sand, earth, and gravel, coated with cinders. The railway triumphed, towns grew all round, and the rest of the Moss yielded to drainage. The subsequent story of the farm at Allwood shows how the industrial and urban revolution helped in its own odd ways to create a new farm in the English landscape.

*

ALLWOOD HOUSE, ASTLEY
by Joan Grundy

Allwood House, Astley, is on the edge of Chat Moss, the great peat bog lying along the northern bank of the Mersey-Irwell, west of Manchester. It used to be a number of small farms. A glance at the modern map is enough to show you that the district was once isolated. The strangling network of built-up land, linked by roads, railways and canals, the pattern typical of south Lancashire, avoids the moss. Only Stephenson's railway and Rindle Road, a mere cart track, cross it, and both were built after 1800.

Before this, Chat Moss was the impassable spongy morass it had been for centuries, impeding communication and threatening farm land with its periodic 'brasting', or overflowing of the moss. In wet seasons it became so swollen with water that huge quantities of black peat flowed out, more than Moss Brook could carry away.

Although local people probably had some paths threading across the bog, in summer the mosses were practically impassable, and far too spongy even for rough grazing. Winter was the best time to get about, when the mosses were frozen hard and the great sheets of flood water iced over, allowing people to flash about on skates. This tradition survived until recently—Fred Green, who farmed at Sales (now part of Allwood House), used to talk of skating down Moss Brook to Rixton about 1900 to 1910.

The narrow, regular shaped fields south of Moss Lane are on reclaimed mossland while the rest of the farm, on the banks of Moss Brook around its junction with Towns Brook, has sticky clay soil. The mossland has been farmed for little more than 100 years but the clay land north of Moss Brook has been cultivated for centuries and a document survives which gives us an idea when it was first cleared.

About the year 1210, Hugh of Astley gave to Cockersand Abbey land in Astley where Moss Brook (called the Fleet in those days) is joined by Towns Brook. 'Riddings' or clearings are mentioned, and 'pannage', which was woodland where pigs could forage.

As well as field shapes, building styles vary with the soils. The old farmhouses and barns on the clay land are long, low buildings of local dark red brick with flagged roofs. Allwood House, built in 1844, stands on the moss. It is a tall, symmetrical building with dining-room and drawing-room on either side of the front hall, kitchen at the back. There is a curious 'missing quarter' at the back where a scullery has been added later. The house is constructed of paler brick with a blue slate roof and the coachhouse block (1869) is of similar materials. All the houses formerly drew their water from wells but

128

15. ALLWOOD HOUSE, ASTLEY

16. Brick House Farm, Rowley Regis

Allwood House also had a vast rainwater tank which was accommodated in the 'missing quarter'.

Moss Brook must have been the edge of cultivation for generations, the land to the south of it being used as common when it was dry enough to carry stock. In 1764 this land, known as Astley Common, was enclosed and with it part of Chat Moss. The allotments made at this time can still be traced on the ground, but enclosure did not

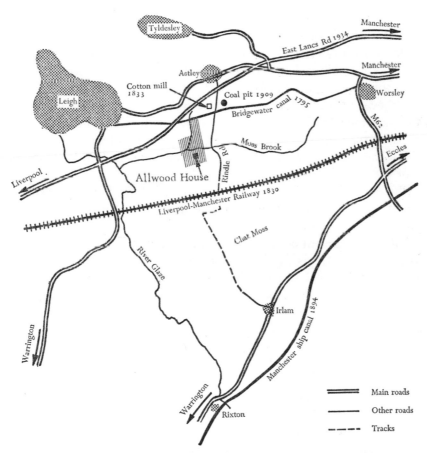

FIG. 11. Allwood House, Astley (Lancashire): general site

necessarily mean cultivation. A map of 1818 indicates small areas of reclaimed land south of Moss Lane but the scale is too small to identify them satisfactorily. As late as the tithe map of 1845 much of the mossland was still uncultivated and even today raw moss, which is cut for moss litter, forms the southern boundary of the farm.

In 1845 thirteen people worked scattered fields all over the present farm and there were eleven houses or cottages. Most of these dwellings have now vanished and only two are occupied today. These fragmented holdings had become five reasonably consolidated farms 100 years later and the final amalgamation has taken place over the last 20 years.

The mushrooming of population which accompanied the Industrial Revolution made land close to the growing towns valuable for food production. This provided an impetus to drainage which started seriously in 1805 with the cutting of ditches and the making of a road across the moss for the first time. So that they would not sink into the bog, the men doing this early drainage work wore flat wooden pattens strapped to their soles.

The Stephensons' railway was laid with some difficulty across the moss between 1826 and 1830 and this gave a further boost to the work of drainage.

When the land was dry enough to carry horses wearing pattens it was ploughed and the tough, wiry sod usually raked up and burned. Now the soft spongy moss had to be built up into a soil, and scores of acres were covered with marl in the old manner. Much of the new land near the railway was marled from the pit at Sales Farm. A tramway was laid from the pit, along Turf Lane, and across the raw bog, and horses hauled trains of about seven little trucks loaded with marl. The pit itself was formed in a series of steps, and forkfuls of the sticky clay were thrown up from one step to the next, then loaded into trucks or carts.

Green's, who farmed Sales Farm for many years, used marl on their fields until about 1900. The land was laid out in 4-yd. butts and marl was dumped in heaps down the middle of the butt and left to weather before spreading.

Also used in huge quantities for building up the moss was nightsoil, available in profusion from Manchester as well as from the privies of local towns. Some idea of the embarrassing quantities available can be gained from the fact that the population of Manchester and Salford trebled between 1801 and 1851, and had doubled again by 1901.

Astley farmers were lucky in one respect, they need take only as much as they required—too much nightsoil was found to be harmful and nothing would grow, but Irlam Moss, south of the railway, belonged to Manchester Corporation and farmers there had to take it as fast as it came. Special nightsoil trains came from Manchester every day along a light railway and tramlines were laid across the moss.

On Astley Moss there was a farmers' siding where nightsoil was collected by the farmers themselves. It also came by canal to the manure wharf at Astley where neighbours with their carts would load it together as it had to be moved within a certain number of days. As well as nightsoil, slaughterhouse waste and manure from town stables and dairies was collected from the 'muck-wharf'.

In spite of the drainage and manuring the moss remained soft and spongy to the tread and the horses had to wear pattens. On some farms these were worn on all four feet but here only on the hind feet. The horses needed time to get used to their pattens, square or oval, about 10 in × 12 in., and clipped on to the hoof. They used to cut the inside of the opposite leg, and they made it difficult for the horses to dig their toes in for a hard pull. Pattens lamed many a good horse.

The moss fields are still spongy, especially in wet weather, and there is always the likelihood of getting bogged with heavy equipment.

As the moss dried out and cultivation proceeded, new houses began to spring up. One of these was Allwood House, built in 1844 by John Allwood, one of a family of Manchester wheelwrights. Only the long enclosure in which it stands belonged to the house at that time. After his death another Manchester businessman called Heywood lived there and in 1869 built the stable and coachhouse block.

The moss soils lent themselves to market gardening and produce found a ready market in the growing towns. Celery was grown at Allwood House until 1920. Dairy

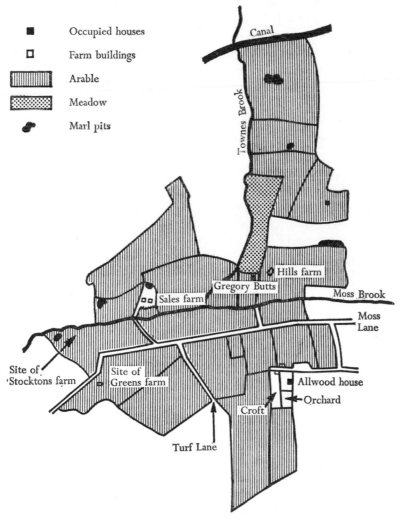

FIG. 12. Allwood House, Astley: plan of farm today

131

produce was also in heavy demand, and the Greens, at Sales Farm, had their shippons full of cows. They made butter which they carried to the markets at Manchester or, more often, Warrington, and the skim was fed to the calves and pigs. There are no stock at all on the farm now.

There was also keen demand for moss litter in the horse transport age, as it is about three times as absorbent as straw. Large areas of moss were let to Moss Litter Companies which are still active.

When the tithe map was drawn in 1845 a large proportion of meadowland was recorded, showing that hay was already an important crop in the district. Manchester had become a busy trading centre and thousands of horses were stabled in the city, as well as dairy cattle, and mossland farmers were not slow to take advantage of this vast market within such easy reach.

The meadows were well looked after, they were heavily manured and until about 50 years ago were irrigated. The main ditches had floodgates, called clowes, at their outfalls into Moss Brook and when the brook rose in spring they were opened to allow water on to the meadows. Usually the water was allowed to flow up the reans (furrows) but sometimes a furrow would be ploughed out as a gutter. Flooding left a fine layer of silt over the field, and the extra water was beneficial in those days of better weather.

People were so keen on irrigation that if they heard rain during the night they would get up to open the clowes though during haytime it was more important to get up and close them or the rising water would leave silt in the hay.

When the time came for the hay to be sold it was cut out of the stack in trusses weighing anything over 56 lb. This was a skilled job and if you could not do your own, professional hay cutters would come. The trusses were about 3 ft. 6 in. × 2 ft. × 2 ft., and were tied with straw bands.

It took about $3\frac{1}{2}$ hours to walk to Manchester with a load of hay, so this meant an early start. Many farmers went to the haymarket in the old part of Manchester around Shudehill and waited for buyers, but it was better to have a customer like the Pendleton Carriage Company where you could deliver the hay and return with a load of manure without hanging about.

The coming of motor transport ended this trade, it was quicker for merchants to collect hay at the farm, and also motor vans and lorries gradually ousted town horses and later country ones. At Allwood the first tractor was bought in 1939 but two horses were kept until 1948.

A Black Country Farm

The Black Country is now one of the most ravaged parts of England, though right down to the beginning of the nineteenth century much of it remained rural. Large parts of it were heathland, not of much agricultural value though useful, indeed essential, for providing common grazing for all the farms round about. As late as 1806 it is said that the greater part of where the conurbation of West Bromwich now stands was nothing but a rabbit warren. Numerous place-names testify to the fact that heathland covered a very large proportion of this now-derelict countryside.

Professor Wise's account of the history of Brick House Farm is a classic study of how the Industrial Revolution overwhelmed one of these old farms step by step. Today nothing can be seen of its buildings, and its former fields became a mess of abandoned coal-tips, quarries, clay-pits, old furnaces, and all the debris of an exhausted industrial landscape. This in turn has been largely cleaned up, and a council housing estate now completes the history of this fascinating bit of English landscape.

This study is based upon the records of the trustees of Deritend Chapel in Birmingham, now housed in the Birmingham Reference Library. The trustees acquired the farm, together with other land, in 1677 for the maintenance of the chaplain and, in default of other funds, the repair of the chapel. It was never much of a farm, but unknown to anyone valuable seams of coal and ironstone lay 100 yds. below the surface.

In the early years of the nineteenth century, especially, the South Staffordshire coalfield developed rapidly, and the trustees of the chapel in Birmingham found themselves sitting, not on a gold-mine but on something just as good. In the year 1821 they took powers through a special Act of Parliament to grant mining leases. Three years later mining began beneath Brick House Farm: the following account shows precisely what happened after that momentous event.

*

BRICK HOUSE FARM, ROWLEY REGIS (STAFFS.)
by M. J. Wise

Interspersed among the factories, foundries, and forges of the Black Country, the industrial complex in the heart of the English Midlands, the modern observer may find patches of grazing-land and even small fields, which serve as a reminder of the former rural state of this area. Two hundred years ago this was a rural area; against the background of farms, fields, and heathland, an industrial pattern was only beginning to emerge. Towards the end of the eighteenth century the conversion of land from farmland to industrial use became more common. Canals cut farms in half, blast furnaces sprang up, and coal mines developed in fields which, a few months before, had borne their last crops. The coming of the railway in the mid-nineteenth century assisted in the transformation of the scene from an agricultural to an industrial one.

It has been found possible to study the history of a small farm which was engulfed in the industrial development of the nineteenth century and to trace the successive stages in the decline of farming standards and of farm management for which the increasing concentration of industry and of the industrial population was responsible. Particular emphasis has been given to this by recent interest in the lowering of farm standards in Britain caused by extensions of urban and industrial areas.

The farm in question was the Brick House Farm, which lay in the Rowley Regis district of South Staffordshire some 2 miles south of Dudley and $6\frac{1}{2}$ miles north of Birmingham. The farm formed part of the property of the Trustees of the Estates of St. John's Chapel, Deritend, Birmingham, from whose minute books the successive stages of decline can be traced. Favourably situated, with a western aspect, on the slopes of the Rowley Hills, a prominent local feature, the farm possessed, in common with other farms in the district, heavy soils derived from the local Etruria Marl, a heavy red-purplish clay belonging to the Upper Coal Measure series. A hundred yards below the surface, however, lay a number of valuable seams of coal and ironstone. Of especial importance was the presence of the celebrated 'Thick' or 'Thirty-Foot Seam', a seam of high quality and extraordinary thickness, which was highly sought after and was responsible, more than any other single factor, for the early and successful exploitation of the South Staffordshire coalfield.

Farming in this region seems never to have been of an outstandingly high quality, largely no doubt because of the heavy nature of the soils. However, during the first quarter of the century the state of Brick House Farm seems to have been of an average standard for the district. The farm comprised some 66 acres, valued in 1830 at a rent of £3 an acre. The leases contained numerous covenants designed to secure a good standard of maintenance with provision for regular manuring and, especially important in this district of clays and heavy soils, adequate drainage. In 1831 the new tenant was required 'to drain by furrow soughing at least 10 acres per annum for the first four years'. Each field when in fallow was 'to be gathered into lands of 5 yards width to facilitate the drainage'.

During the early years of the nineteenth century, this district had not been seriously affected by that colliery and industrial development which was rapidly transforming the landscape of other parts of the Black Country. In 1824, however, the mining of coal beneath the Brick House Farm began, reflecting, incidentally, the current spread of mining into the Stour valley sector of the coalfield, in which this farm was situated.

Before many years had elapsed, the results of the growth in intensity of mining and of the increase in population of the district had become apparent in the condition of the farm. The reports of visiting Trustees in 1843 and subsequent years make clear the trend of events. Damage to the farm was caused chiefly through mining operations. The subsidence of the surface hollowed and pitted the land, breaking drains, altering watercourses, and making field drainage difficult. In consequence, rushes and sour grasses grew extensively. Spoil-banks ruined the value of the land they occupied and cinders and waste were washed and carried over wider areas, increasing the damage. Though some rebates of rent were made to the tenant in these respects, they did not fully offset the total damage and dislocation caused. Neighbouring industrial and quarrying activities were a further source of nuisance, for spoil from the nearby dolerite quarries extended over some parts of the farm. With the general growth of population, the incidence of trespass greatly increased the damage to crops, gates, and fences. In consequence of these disadvantages a Birmingham surveyor assessed the value of the farm at only £150 a year but the tenant in 1847, was, in fact, paying £180.

By 1858 the land was described as 'in a bad state and excessively poor. From sinking of the surface throughout the colliery, the drains are broken'. Subsidence was pronounced and the pitbanks were an increasing obstacle to good farming. In 1865 'crownings in', or sinking of the surface around the mouths of old shafts, had rendered some areas useless. Conditions were not such as to encourage good management and difficulty seems to have been experienced in obtaining able and whole-time farmers as tenants. One by-product of the mines was the availability of burnt spoil from pit heaps which had been on fire, and this was said to be a useful fertilizer, large quantities being quarried for that purpose. Attempts were made, with some success at first, to grow early potatoes for the local market.* But tenants had little capital to spend on improvements and there was a general reluctance to spend large sums of money in reclaiming land which was still liable to subsidence and around which signs of industrial growth were increasing year by year. Some areas became infested with a 'noxious weed, red leg'. By 1868 as was perhaps to be expected, the quality of management had deteriorated sadly and thistles and weeds bore 'silent but forcible testimony to the want of efficient farming'. One Josiah Lee had held the farm from 1858–69, of whom it was reported that the land which was 'in a most impoverished condition when he took it' was 'if possible, in a worse condition when he left it'.

The new tenant, William Davis, was only a part-time farmer. He had a supplementary occupation in a local iron-works but he began at a rent now reduced to £100 a year as though he meant to rectify the situation. It was not long, however, before his available capital was well-nigh exhausted and much sterility and waste remained. Trespassing had now become a most serious problem and legal action in the County Courts proved

* This was not unusual and other references exist to the practice of growing 'early potatoes for the London market raised in ground near Dudley, heated by steam and smoke which proceed from an old colliery which has been on fire for some years.' Parliamentary Papers, *Midland Mining Commission First Report* (S. Staffordshire, 1843), v.

only a partial and expensive answer. The Trustees and their tenant were battling against the whole course of industrial development of the Black Country, for mines, quarries, ironworks and firebrick works with their concomitant workers' settlements were growing steadily in this neighbourhood. The reports of the Trustees make no mention of atmospheric pollution but there is ample evidence of the result of this on farm crops and on the natural vegetation from other parts of the Black Country.* Furthermore, in conditions which would have been marginal in prosperous times, the great Agricultural Depression had also to be faced.

For a time, an experiment in sheep and poultry farming, with 60–80 sheep, was carried on, but conditions proved hardly suitable for this purpose especially because of the lack of drainage in most parts of the farm. Crop production, which was only moderate in good years, was negligible in poor ones, and cadlock† grew in quantity among the

BRICKHOUSE FARM AND ITS VICINITY c.1880

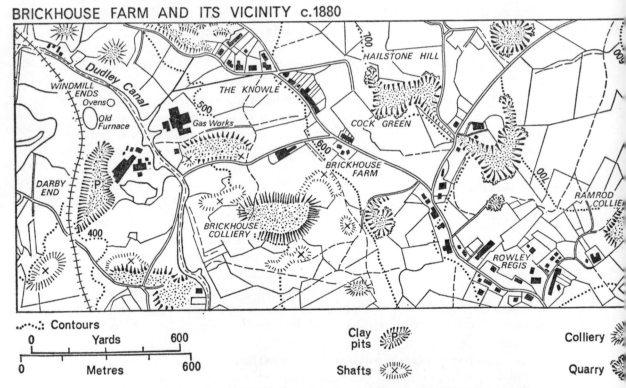

FIG. 13. Brick House Farm, Rowley Regis (Staffordshire) *c.* 1880

* The situation in the nineteenth century was well described by the local historian F. W. Hackwood, *Odd Chapters in the History of Wednesbury* (1920), 65. '. . . pit spoil mounds and cinder heaps: with gloomy smoke laden skies and an atmosphere so heavily charged with sulphur and other deleterious exhalations from its myriad groups of chimney stacks that vegetation shrivels, fades slowly and perishes of blight.'

† Charlock.

barley and wheat. Gates and fences were in bad repair, and subject to constant damage. The Reports of visiting Trustees make increasingly depressing reading. In 1879 ditches were choked with pit refuse and dust and ashes were encroaching on to the farm. Matters were not helped by 'the untidy farming habits of the tenant', which seem to have been responsible for the 'slovenly and untidy appearance of the farm'. Parts of fields separated by spoilbanks from the main farm had become difficult of access and of little use.

By 1885 the problem had become acute. The farm 'still bore the poor and desolate appearance', (this in midsummer) 'that it had on the occasion of our last visit'. Rain came in, on occasion, through the farmhouse roof, crops were poor and largely inter-mingled with cadlock. The Trustees were forced again to spend considerable sums on repairs, maintenance and drainage and, with the coal beneath almost worked out, must have considered the farm to be, by this time, almost as much a liability as an asset. They had, however, derived a considerable income from coal royalties. The tenant in 1885 had given notice to quit and was prepared to continue only on condition that the rent was reduced to £75 a year. But in the view of a local observer considering its 'impoverished state and the general state of agriculture' the farm was 'worth not more than £50 a year'—well under £1 an acre.

In 1886 the Trustees came reluctantly to the conclusion that better returns could be obtained from building than from farm leases and parts of the farm flanking the roads of the district were advertised as to let on building leases. Some cottages were, in fact, erected, but at a time when the final engulfment of farming by building and industry seemed imminent, a temporary reprieve was granted. Due to the twin causes of exhaustion and flooding of the coal seams, mining activity declined and finally ceased. Neither was there any further large expansion of iron working in the immediate vicinity and the rate of increase of population slackened. So the farm was carried on, managed by men who were part-time farmers, but engaged primarily in the local iron and general metal industries. Surrounded on almost all sides by quarries, ironworks, firebrick works, and other evidences of the industrial age, the standard of farming was not high. More and more houses were built around the farm boundaries. But, stimulated by the demand for food in World War I, agriculture survived for a time.

In 1950 the visitor to Brick House Farm found the farm buildings a mere ruin. The fields and meadows which were tilled and maintained with such pride more than a century ago were covered with rank grass and reeds while above them stood the abandoned colliery pit heaps. Much of the farm could not be seen even in this state, for a modern housing estate, erected by the local council, had obliterated even the slightest traces of its former rural character.

This is a story which could be repeated for almost all parts of the Black Country and for many other industrial regions of Britain as well. To take but one example, let us notice the case of the village of Ettingshall, a few miles north-west of Rowley. A report on this village in the early 1840s recorded a diminution of the tithes payable to the rector by nearly one-fourth in seven years for the land was now 'so much broken up by being undermined and covered with pit banks and ironworks and buildings'.* This district had been, thirty years earlier, almost completely agricultural with 'hardly a house . . .'. But 'the cultivation is transferred from the surface to the interior of the earth and this latter requires many more labourers than the cultivation of the surface and produces much more wealth'.

The Black Country today bears many acres of waste land, scarred by pit banks and slagheaps, which were once farmland with histories similar to that of Brick House Farm. Today the grazing of animals and the occasional allotment are, for the most part, the limits of farming activity. But it is still possible to see farming in decline in many areas on the fringe of growing towns and industrial areas, and it is pertinent to wonder whether, in fact, the lessons of Brick House Farm have yet been fully learnt.

* Parliamentary Papers, *Midland Mining Commission, First Report*, 50.

Index

INDEX

Scotland, 11, 17, 18, 25, 111–23
Seccombe family, 103
Sherborne Abbey (Dorset), 51, 53
shippon, 37, 40, 41, 43, 44, 45, 46, 85
Shirley family, 56
soils, 17, 18, 31, 42, 52, 57–8, 86, 92, 121, 131, 133, 134
Somerset levels, 21
South Tawton (Devon), 42
Spitchwick (Devon), 41, 42, 45
Stanchester (Dorset), 51
Stephenson, George and Robert, 127, 130
stone-dykes, cost of, 117–18
Summerbank Farm (Caithness), 11, 111, 119–23, Plate 14

Talbot family, 56, 57
tenancies, length of, 57
Thames, 20
Truro Museum, 30
Tudor Cornwall, 30

villages, 15–16

wages, 70, 89
Wales, 11, 18, 19, 29
Walker family, 93–4
Walton, James, 85
water-supply, 16–17, 29, 30, 37, 40, 43, 57, 58, 59, 65, 80, 86–7, 92, 114–15, 132
Wednesbury, 136 n.
Welland, 20
Wells family, 56
Welsh borderland, 17
West Bromwich, 133
Whiddon Down (Devon), 48
Widecombe-in-the-Moor (Devon), 42
Wilstrop family, 91, 96–7
Wilstrop Hall (Yorks), 21, 90, 95–9, Plate 10
Wise, Michael, 11
women, influence of, 23–4
Wrixhill Farm (Devon), 36

yeoman pedigrees, 103
Yorkshire, 18, 20, 21, 79–99

141